T0224761

Zircon, Zirconium, Zirconia - Similar Names, Different Materials

Bożena Arnold

Zircon, Zirconium, Zirconia - Similar Names, Different Materials

Bożena Arnold
Waldbronn, Germany

ISBN 978-3-662-64268-9   ISBN 978-3-662-64269-6   (eBook)
https://doi.org/10.1007/978-3-662-64269-6

The translation was done with the help of artificial intelligence (machine translation by the service DeepL. com). A subsequent human revision was done primarily in terms of content.

This Springer imprint is published by the registered company Springer-Verlag GmbH, DE, part of Springer Nature.
The registered company address is: Heidelberger Platz 3, 14197 Berlin, Germany

Preface

Minerals can charm us as beautiful gemstones and serve as valuable materials. Their hardness, optical properties, chemical and thermal resistance, and other special characteristics are irreplaceable for certain applications. All this fascinates me about these materials, which can be of natural origin or artificially produced.

The idea of the book started with my interest in a diamond-like gemstone called "zircon." As a mineral it belongs to the group of silicates and is of great importance for geochronometry because of its possible great age.

Its competitor as a gemstone bears the name "zirconia." It does not occur in nature in this form, and from a material-technical point of view, it belongs to the group of oxides.

The metal "zirconium" plays an important role in nuclear technology and its oxide forms an important group of ceramic materials.

The names mentioned are amazingly similar and confusion is almost inevitable, which in turn often leads to erroneous information. I found this exciting, but also irritating. So I decided to write about it, to contribute something to the distinction of these materials.

Naming is often an adventure. With minerals, it is usually colors to which their names refer. So it is with zircon; its name comes from an old word for gold-colored. The names "zirconium" as well as "zirconia" were derived from "zircon," so the confusion with the names was pre-programmed. And to this day we have to deal with it, especially in medical technology.

This book is a popular scientific treatise on the basic element zirconium and on materials based on it. It is aimed at all those who have a basic knowledge of chemistry and technology and who are interested in various materials without seeking in-depth scientific knowledge.

Waldbronn, Germany
May 2019

Bożena Arnold

Contents

The Constant Confusion: An Introduction

"Zircon", "zirconium", "zirconia"—the similarity of the three names leads one to believe that they are one and the same material. This means, then, that the three terms are used as synonyms. However, from a material point of view, this is not true. One can almost call it a misfortune that the three names are so similar and as a result these materials are often confused.

In this book, the materials addressed (Fig. 1.1) are quite clearly distinguished and at the same time their commonalities are shown.

In order to establish clarity of the terms, right at the beginning you will find information about in the following chapters:

- the mineral known since ancient times called "zircon", which is chemically zirconium silicate ($ZrSiO_4$),
- the hardly known 40th chemical element "zirconium" (Zr),
- the zirconium oxide (ZrO_2), which is one of the most important ceramic materials today and is called "zirconia" as a grown single crystal.

Zircon is one of the most abundant and oldest components of the earth's crust. Because of its similarity to diamond, it is used as a gemstone in the jewellery industry (Fig. 1.1). However, zircon is also the starting material for the production of the pure metal zirconium.

The metallic zirconium was found in zircon and is steadily gaining in importance. The metal is relatively soft, pliable, silvery shiny and corrosion resistant (Fig. 1.1b). Its most important application is in nuclear technology.

The diverse property profile of zirconium oxide enables various applications, for example for kitchen knives (Fig. 1.1c), but also in medical technology. Synthetic single crystals can be produced from the zirconium oxide (Fig. 1.1d), which are inexpensive diamond imitations.

Due to the naming, the confusion is actually as good as pre-programmed. If we look at the chemical compositions, however, we can clearly see that we are dealing

© Springer-Verlag GmbH Germany, part of Springer Nature 2022
B. Arnold, *Zircon, Zirconium, Zirconia - Similar Names, Different Materials*,
https://doi.org/10.1007/978-3-662-64269-6_1

Fig. 1.1 The three materials. (**a**) Brilliant-cut zircon (With the kind permission of CARAT-Edelsteinhandel, Vienna); (**b**) Metallic zirconium (With the kind permission of Graz-Consulting, Prinzerdorf); (**c**) zirconium oxide ceramic knife (With the kind permission of Kyocera Feinceramics GmbH, Neuss); (**d**) Cut zirconia single crystal

here with three different materials that also have different properties and areas of application and must be kept apart.

Correctly, the Duden for the German-speaking world distinguishes the three terms in question from each other and defines them correctly. In doing so, reference is made to the—linguistically speaking—common part of speech of the terms: "zircon". This is possibly the reason why exactly this term is used in various combinations, but often incorrectly.

When searching for information on the Internet, all sorts of things from this area are mentioned and displayed under "zircon". Thus, it is sometimes quite difficult to determine which of the mentioned materials is meant in a specific case. At a dentist you get a zircon crown. In a lambda probe, which is used in every car today, one of the components is often called zircon. And in both cases this name is used incorrectly, because in each case it is not the zirconium silicate, but the zirconium oxide. Even in a dissertation, "zircon" appears in the title, but it is immediately apparent from reading the text that zirconium oxide is being reported. In the list of abbreviations of the thesis, however, zircon is again correctly listed as zirconium silicate. In the jewelry trade, stones are still arbitrarily called "zircon" or "zirconia" (despite a price difference). The name zircon is also not infrequently used for the metallic zirconium.

But you can also find a warning on the Internet that reads succinctly in English: "Not to be confused with zircon, zirconia or zirconium". (https://en.wikipedia.org/wiki/Cubic_zirconia).

The three materials mentioned have the basic element zirconium in common. From this point of view, this book should actually also begin with zirconium. However, based on the history of the materials, we will start with the mineral zircon, because it was known to us first—as a beautiful gemstone, but without knowing what the material was. Then we discuss zirconium, a metal discovered in this very mineral zircon. Only then do we turn to the most important of the three materials discussed, namely zirconium oxide. It also occurs as a mineral in nature and is then called baddeleyite. It was not until the 20th century that zirconium oxide began to be produced synthetically and used as a ceramic material. Today, zirconium oxide, along with aluminum oxide, is valued as an important modern advanced ceramic. Finally, we will deal with zirconia, a single crystal grown from the oxide.

If this book succeeds in arousing interest in the various zirconium-based materials and in imparting some knowledge about them, then it will have fulfilled its intended purpose.

Zircon: A Common Mineral

2

Zircon is a natural mineral that was already known as a gemstone in ancient times and was valued for its beauty—as with almost all gemstones because of its brilliance and color. Zircon is not only one of the oldest known minerals, but also one of the most abundant minerals in the earth's crust.

2.1   From the History of the Zircon

The name "zircon" is associated with the Arabic word "zargun", which means golden. In earlier times, the mineral was often called "hyacinth", which was a flowery allusion to its color. However, even then, colorless zircon crystals in particular were prized and used as imitation diamonds.

The mineral was first named "zircon" in 1783 by the German geologist Abraham G. Werner. His student Christian A. S. Hoffman included zircon in his "Handbuch der Mineralogie" (Handbook of Mineralogy), which he wrote after Werner's lectures. A few years later, the famous German chemist Martin Heinrich Klaproth analysed yellow-green and reddish zircons from Ceylon (today Sri Lanka) and discovered in them "a hitherto unknown, independent, simple earth", to which he gave the name "zircon earth" (Terra circonia). Klaproth then found the same earth in a hyacinth, which is a variety of zircon, whereby zircon on the one hand and hyacinth on the other proved to be, in his opinion, "two species or genera of a peculiar rock family". Only René-Just Haüy united hyacinth and zircon to one single mineral after an exact determination of the crystal forms. The analysis carried out by Klaproth is dealt with again in Sect. 6.1 in Chap. 6, wherein the discovery of the metal zirconium is discussed.

Today we know that zircon is a silicate, more precisely the zirconium silicate with the chemical molecular formula $ZrSiO_4$. Silicates and also quartz (silicon oxide) have an outstanding geological importance. They form about 90% of the earth's crust and about 99% of the earth's mantle. So literally the ground on which we stand and walk is made of these minerals. The moon is also made up of silicates.

© Springer-Verlag GmbH Germany, part of Springer Nature 2022
B. Arnold, *Zircon, Zirconium, Zirconia - Similar Names, Different Materials*,
https://doi.org/10.1007/978-3-662-64269-6_2

According to the frequency of silicates and of quartz, silicon and oxygen are the most common elements of the earth's crust.

2.2   Properties of Zircon

From a chemical point of view, zircon, i.e. the zirconium silicate, can be seen as a combination of 67.2% zirconium oxide (ZrO_2) and 32.8% silicon oxide (SiO_2). The three elements zirconium, silicon and oxygen form a common tetragonal crystal lattice. With a Mohs hardness of 7.5 zircon belongs to the hard minerals and with a density of 4.7 g/cm^3 to the heavy minerals. A high melting point of 2420 °C gives the mineral a very good heat resistance. Zircon is non-magnetic and electrically non-conductive. It is chemically very resistant and insoluble in water, acids, alkalis and also in aqua regia. Hot and concentrated hydrofluoric acid attacks zircon only weakly. Its resistance even exceeds that of diamond.

The great resistance of zircon is actually a surprise because zircon, like another important mineral called olivine, is an insular silicate. This means that its SiO_4 tetrahedra are not linked together to form chains or lattices, but are held together by metal ions. In the case of olivine, these are double positive magnesium and iron ions, while in zirconium, quadruple positive zirconium ions ensure cohesion. This leads to a different crystal lattice in zircon than in olivine, namely the tetragonal one, which is very stable and insensitive to weathering. While olivine weathers easily, zircon remains very resistant.

In addition to the element zirconium, the crystal lattice o zircon always contains the chemically very similar element hafnium (Hf) (Chap. 7). The average content of hafnium oxide (HfO_2) is 0.5–2.0%. Zircons with an increased hafnium content of up to 24% are mineralogically called "alvites". In addition, the rare earth elements cerium (Ce) and yttrium (Y) occur. Zircons often contain the radioactive elements uranium (U) and thorium (Th). The silicates of these two elements have the same structure as the zircon and thus solid solutions of the three minerals can be formed. The radioactive decay of these substituted components causes a progressive destruction of the crystal structure of zircon by emission of Alfa particles, which is called isotropization or metamictization. This phenomenon causes hydration of the mineral, decrease of its density and hardness, and change of other properties. However, it is precisely because of the radioactive elements it contains that zircon is of great interest to science, as its age or the age of rocks containing zircon can be determined on the basis of their decay series. It could be proved that zircons are the oldest minerals on earth with about 4.4 billion years. We dedicate ourselves to this very interesting topic in Chap. 4.

2.3 Occurrence of Zircon

Zircon belongs to the common and rock-forming minerals. Zircons crystallize early at high temperatures from silicate rock melts and are therefore often a component in magmatic granites. Zircon is also found in some metamorphic and sedimentary rocks, which can be traced back to rocks that originally crystallized and were later redeposited. As a very resistant and heavy mineral, it accumulates in sediments and is thus found in secondary deposits, the so-called placer deposits (Chap. 5).

Mostly zircon occurs in the form of opaque and cloudy crystals; transparent crystals are an exception. The pure chemical compound of zirconium silicate is colorless, but certain impurities cause various colorations of zircon. Brown zircons are fairly common, but red stones are rare and sought after. Sometimes zircon forms beautiful colorless and, in addition, translucent crystals that can be mistaken for diamonds and therefore serve as imitation diamonds (Chap. 22). In Fig. 2.1 different coloured zircon crystals are shown. A zircon with the typical brown color, still in the parent rock of feldspar, ouarz, and mica (i.e., granite), is shown in Fig. 2.1a. Figure 2.1b shows two red zircons found in Burma (now Myanmar), where some of the known deposits are located.

Zircon has many varieties, which differ in color, shape of crystals and also in composition. In addition to the two varieties already mentioned, "hyacinth" (yellow and yellow-red to red crystals) and "alvite", a further nineteen varieties of zircon are named and described in mineralogy. Among them are, for example, "ribeirit", an yttrium-rich zircon from Brazil, or "oerstedtite", a metamict zircon from Norway, which was named after the famous Danish physicist Hans Christian Oersted.

Due to the tetragonal crystal lattice, individual zircon crystals can reach considerable size. A crystal weighing about 2 kg was found in Australia and one weighing almost 4 kg in Russia. Mostly, however, zircon is found in the form of tiny crystals, ranging in size from 0.1 to 0.3 mm, even smaller than the head of a pin. As already mentioned, zircon is widespread on earth, but it usually occurs in relatively small quantities in the individual rocks. The worldwide occurrences of zircon are accompanied by a number of other minerals, including quartz, topaz and spinel.

Fig. 2.1 Zircon crystals. (**a**) Brown zircon in the parent rock; (**b**) Red zircons (With the kind permission of Mr. S. Ellenberger, Crystal-Treasure, Kassel).

Findings of the mineral have been documented in Finnmark/Norway, in the Erzgebirge and in the Eifel/Germany, in Australia, in Brazil, in Canada as well as in many other countries.

The main source of zircon, and therefore of all zirconium-based materials, is zircon sand (Chap. 5). Zircon is mined from natural placer deposits, then concentrated by various techniques and further processed into various materials. It is the most important raw material for both zirconium (Sect. 6.2 in Chap. 6) and hafnium (Chap. 7), as well as for zirconium oxide (Chap. 12).

The chemical compound zirconium silicate can be produced by fusing silicon oxide (SiO_2) and zirconium oxide (ZrO_2) in an electric arc furnace or by reacting a zirconium salt with sodium silicate in aqueous solution. However, taking into account the fact that we can find a lot of zircon in the earth's crust, its synthetic industrial production is neither useful nor necessary.

Further Reading

1. Bayer, G., & Wiedemann, H. (1981). Zirkon—vom Edelstein zum mineralischen Rohstoff. https://onlinelibrary.wiley.com/doi/abs/10.1002/ciuz.19810150305. Accessed: 15. Aug. 2018.
2. Elsner, H. (2006). *Bewertungskriterien für Industrieminerale, Steine und Erden, Teil 12: Schwerminerale. Geologisches Jahrbuch Reihe H, Heft 13* (S. 55–69). Hannover: Bundesanstalt für Geowissenschaften und Rohstoffe und Landesamt für Bergbau, Energie und Geologie.
3. Markl, G. (2015). *Minerale und Gesteine* (S. 48). Berlin: Springer Spektrum.
4. Weiß, S. (2011). Seiland. Norwegen—eine legendäre Zirkonfundstelle am Alta-Fjord, Finnmark. *Lapis, 11*, 15–25.
5. Schorn, S. Mineralienatlas–Fossilienatlas. https://www.mineralienatlas.de/lexikon/index.php/MineralData?mineral=Zirkon. Accessed: 11. Juni 2018.
6. Wikipedia. Zirkon. https://de.wikipedia.org/wiki/Zirkon. Accessed: 5. Aug. 2018.

Zircon: A Genuine Gemstone

3

Zircon has all the properties that are asked for in minerals if they are to be used as gemstones. For a mineral to become a valued gemstone, it must have a high hardness (more than 6 on the Mohs scale) and high light refraction, and should also be colorful. Zircon meets these requirements. However, it does not belong to the "Big Four" group, which includes diamond, ruby, sapphire and emerald. The beauty, rarity and permanence of gemstones have always impressed us greatly. In an ever-changing world, only gemstones actually remain unchanged.

The zircon can compete with the most famous gemstone, the diamond. Cut in the brilliant cut, it shows a similar play of light and color as the latter (Fig. 1.1a in Chap. 1). However, despite its almost ideal properties, zircon is fairly unknown as a gemstone in the jewelry field. Zircon crystallizes in a tetragonal lattice. Large crystals of this so-called high zircon form octahedra, which resemble two pyramids joined at the base. To be distinguished from this are so-called deep zircons, which no longer have a crystalline structure as a result of the radioactive decay of uranium and thorium (Sect. 2.2 in Chap. 2)—they are amorphous.

The greatest yield of cut-worthy zircon crystals, which are considered gemstones, is achieved by the classic Asian gemstone countries of Sri Lanka, Myanmar, Cambodia, Thailand and Vietnam. Zircon participates in the development of various minerals. It is almost invariably placer deposits where zircon occurs as an accompanying material to ruby, sapphire and other gemstones. Newly discovered, hopeful zircon deposits are located in Tanzania.

Zircons come in all colors, but they are mainly "warm" shades such as yellow, brown, orange, also red and pink. However, the pure compound, the zirconium silicate, is colorless, because it does not contain any elements that have a color-giving effect. It is highly probable that the zirconium colours are not solely due to the usual colour-bearing elements such as chromium or iron, but result from the addition of rare earth elements such as cerium (Ce) and yttrium (Y). The radioactive elements uranium and thorium mentioned above also cause colour changes; for example, green zircons are formed. The presence of these elements is also the reason for the changeability of zircon colors by heating. The most popular colour variations—blue

© Springer-Verlag GmbH Germany, part of Springer Nature 2022
B. Arnold, *Zircon, Zirconium, Zirconia - Similar Names, Different Materials*,
https://doi.org/10.1007/978-3-662-64269-6_3

Fig. 3.1 Cut zircons in different colours. (Recorded in the German Gemstone Museum in Idar-Oberstein)

zircon, which is very rare in nature, and colourless zircon, which is similar to diamond—are produced by heat. The blue color is produced when the zircon is heated in a vacuum or under reducing conditions. In contrast, heating at temperatures of 850–900 °C in oxygen removes any color and the zircon becomes colorless. The latter variety is very often processed in brilliant cut. Not all heat-treated zircons are colourfast. Under ultraviolet radiation or in daylight, they often take on their original color. Therefore, zircon jewelry should not be exposed to the heat influence of sunlight or strong illumination.

In the zircon showcase in the German Gemstone Museum in Idar-Oberstein, some different coloured zircons can be admired. Figure 3.1 shows a small selection of these cut zircons.

Zircon has very good optical properties, which is advantageous, even decisive, for gemstones. Its light refraction of 1.92–1.99 is high (it is significantly lower in green deep zircons) and it also has—due to its tetragonal crystal structure—a high birefringence of +0.055. Despite the high light refraction, the numerous colours of zircon sometimes do not come out so clearly and brightly because they are dampened by this strong birefringence. Its high dispersion (light scattering) of 0.038 (BG value) gives zircon an intense sparkling fire that is even stronger than that of a diamond. Zircon exhibits pleochroism, that is, it changes color when held at a different angle of illumination. However, this multicolorism is usually only faint. Table 22.1 in Chap. 22 lists the properties of zircon.

As a gemstone, zircon also has some negative sides. It is very brittle and therefore easily chips off at the cut edges. When set, it attracts grease and dirt to the bottom of the stone and can therefore become dull and look unsightly after a certain period of wear. Zircon is not suitable for ultrasonic cleaning, which is often used on gemstones. During repair work, zircon jewellery should not be exposed to heat because, as already mentioned, a change in colour is possible (especially with colourless zircon). Also the natural radioactivity, although low, is seen rather as a disadvantage of zircon.

Zircon was the best natural diamond imitation until the end of the 1970s. In Chap. 22 we deal with the comparison of zircon and diamond. Zircon experienced its heyday in the sixteenth century Europe, when it was often processed by Italian jewelers. Today, the competition of the artificial product "zirconia" (Chap. 21) is the main reason for a lower appreciation of zircon on the jewellery market. However, it remains a genuine and beautiful gemstone.

Further Reading

1. Henn, U. (2013). *Praktische Edelsteinkunde* (S. 148–150). Idar-Oberstein: Deutsche Gemmologische Gesellschaft.
2. Linsell, G. (2013). *Die Welt der Edelsteine* (S. 202–209). Berlin: Juwelo TV Deutschland GmbH.
3. Schumann, W. (2017). *Edelsteine und Schmucksteine* (S. 124). München: BLV Buchverlag.
4. Symes, R. F., & Harding, R. R. (2012). *Edelsteine und Kristalle* (S. 304–305). München: Dorling Kindersley.
5. MediaWiki. Zircon. http://gemologyproject.com/wiki/index.php?title=Zircon. Accessed: 16. Nov. 2018.
6. Rössler, L. Edelstein-Knigge. http://www.beyars.com/edelstein-knigge/lexikon_570.html. Accessed: 20. Nov. 2018.

In the Service of Geology

4

For a long time, zircons have been valued and used for their beauty and similarity to diamonds (Chap. 3). Today, zircons are famous and attract special interest for a completely different reason: A tiny zircon discovered in Western Australia is the oldest mineral known so far in the world, with an age of about 4.4 billion years. Zircon thus occupies a significant position among minerals. This is interesting and fascinating at the same time.

4.1 The Atomic Clock

The commercial value of zircons has decreased over time. On the other hand, their scientific value has increased, especially for geology, more precisely for geochronometry. Zircon crystals help scientists reconstruct the history of our Earth. But why can zircons become so old and how do we know their age?

Zircons are the most stable components of the earth's crust. Unlike many other minerals, they can survive extreme changes in the earth's crust. They practically do not weather, especially not in granitic magma when it cools. Zircons persist even when their parent rock has long since eroded. Even in heavily weathered sand, which consists of about 99% quartz, a certain amount of zircon particles can be found. Precisely because of their resistance, zircons are important witnesses from the early days of the earth. In addition, there is an atomic clock inside them. These special conditions make zircons so valuable for geologists.

Because zirconium atoms are quite large, other large atoms such as uranium and thorium can become incorporated into the crystal lattice during the formation of zircons, which have no place in other minerals (Sect. 2.2 in Chap. 2). Radioactive atoms decay over time to form other elements. For example, the most common uranium isotope, U-238, decays to lead (Pb) with a half-life of about 4.5 billion years. The decay of uranium was used to date rocks as early as 1907, shortly after the discovery of radioactivity. The uranium-lead system is still one of the most accurate and common methods of age determination in geology. A prerequisite for its use is

© Springer-Verlag GmbH Germany, part of Springer Nature 2022
B. Arnold, *Zircon, Zirconium, Zirconia - Similar Names, Different Materials*,
https://doi.org/10.1007/978-3-662-64269-6_4

that the mineral to be analyzed should contain little or no lead to begin with. This precondition is fulfilled for some minerals important for dating, especially for zircon, baddeleyite and monazite. In the formation of zircon, no lead at all is incorporated into the crystal because lead atoms do not fit well into its crystal structure. So the more lead atoms you find in the zircon crystal compared to the uranium atoms, the older it is. In short, the decay of radioactive uranium into stable lead atoms can be used to determine the age of rock samples. There are few naturally occurring radioactive elements that decay so slowly that they are suitable for dating very old rocks. This built-in radioactive (equal to atomic) clock makes zircon a unique mineral.

4.2   Geochronometry with the Aid of Zircon

Geochronometry deals with the age determination of rocks and minerals by means of time-dependent physical and/or chemical processes. The most commonly used process is radioactivity.

For such dating suitable and special methods as well as equipment are used. Each method begins with the preparation of samples. In the case of zircon, it is particularly complex. First of all, the resistant zircons have to be extracted from granite rocks by force. To do this, the rock fragments are ground into a fine powder. This is then mixed with hydrofluoric acid, one of the most corrosive acids of all. It dissolves almost every mineral – except zircon. After the acid bath, only the zircon crystals remain, which can now be analyzed.

In geochronometry, mass spectrometry methods are the preferred analysis techniques. The principle of a mass spectrometer is based on the fact that the ions arriving in it are deflected by applied magnetic fields and separated from each other. The substances to be investigated are ionized in the gas phase in a high vacuum. Isotope ratios of a chemical element are always determined, never element concentrations. Today, several mass spectrometry methods are available for highly complex isotope analysis. The dating of zircons is mainly done with the method of the "Sensitive High Resolution Ion Micro Probe". This name gives a nice abbreviation: SHRIMP, which is the most commonly used name for the method of analysis. In this method, one uses a high-energy ion beam of oxygen and argon ions to produce secondary ions, which are then separated by their mass and energy in a mass spectrometer and counted with a counter. The SHRIMP microprobe can be used to measure the isotope and element distribution in very small areas down to a size of 10 µm. Therefore, it is also well suited for the analysis of small amounts of complexly structured minerals. Different zones can be analysed separately. SHRIMP analysis is performed on samples in the solid state, no homogenization and transfer into solution is necessary. The most common application is radiometric dating using the uranium-lead method, although the SHRIMP method can also be used for the analysis of other isotopes and elements.

The main supplier of SHRIMP instruments is the company Australian Scientific Instruments, which remarkably fits to the findings of the oldest zircons just in

Fig. 4.1 Geochronometry. (**a**) SHRIMP device (© Geoscience Australia CC BY 4.0); (**b**) Ancient zircon crystal (© Geoscience Australia CC BY 4.0)

Australia. There are about 200 such instruments available worldwide today (2019). The majority are located in Asia and Australia, i.e. in regions where there are large deposits of minerals and gemstones. The nearest device for Germany is located in the Państwowy Instytut Geologiczny (Institute of Geology) in Warsaw/Poland. SHRIMP devices are very large and require special premises (Fig. 4.1a).

4.3 The Oldest Zircons

In Western Australia, about 800 km north of Perth, lies a range of hills called the Jack Hills. The hills became famous because geologists found a handful of zircons embedded in relatively young sandstone. Subsequent dating of the tiny crystals using the uranium-lead method revealed an age of up to 4.404 billion years. These zircons are just a little younger than our planet. That's pretty impressive considering the Earth was formed less than 150 million years earlier. It is widely believed that it had no solid crust to begin with.

Figure 4.1b shows an old zircon. The small crystal was photographed in transmitted light and consists of a core surrounded by layers of different geological ages.

In addition to the oldest dating so far, the crystals from Jack Hills provided further surprising findings. Analysis of gas bubbles trapped in the zircons revealed evidence of the composition of the Earth's atmosphere more than four billion years ago. The mineral grains already show an isotope of oxygen, indicating weathering involving water. They indicate that the Earth already had a thin crust at that time and oceans covered its surface. Oceans are therefore old. Overall, the rock dating and the analysis of enclosed gases show that the young Earth cooled rapidly.

Furthermore, the Western Australian zircons caused an additional stir when traces of graphite were found in them. This could be of biological origin. This is suggested by the ratio of stable carbon isotopes in the graphite, which corresponds to the isotopic signature of living organisms. The zircons studied were younger than those that suggested liquid water on the early Earth. According to the latest data, life may be as old as the first oceans on the cool Earth.

The oldest mineral in Europe is considered to be a zircon, which is 3.69 billion years old and was found in gneiss rock in northern Norway not far from the town of Kirkenes. The oldest mineral in our solar system so far has been found in lunar rock called breccia 72215. And it's a zircon again, at 4.417 billion years old, only 100 million years younger than the Earth's satellite itself. It thus surpasses in age the earliest zircons in Earth's history from Australia's Jack Hills.

From a geological point of view, we owe many interesting findings to zircon.

Further Reading

1. Feil, S., Resag, J., & Riebe, K. (2017). *Faszinierende Chemi* (S. 80–81). Berlin: Springer.
2. Krzeminska, E. (2014). Mikrosonda jonowa SHRIMP IIe/MC. *Przeglad Geologiczny, 7*, 373–374.
3. Markl, G. (2015). *Minerale und Gesteine* (S. 544–552). Berlin: Springer Spektrum.
4. Neukirchen, F. (2012). *Edelsteine – brillante Zeugen für die Erforschung der Erde* (S. 143–149). Berlin: Springer Spektrum.
5. Prothero, D. (2018). Zirkone – Zeugen der frühen Erdgeschichte. *Spektrum der Wissenschaft, 9*, 56–60.
6. Sci-News. http://www.sci-news.com/geology/science-jack-hills-zircon-oldest-known-fragment-earth-01779.html. Accessed: 5. Sept. 2018.
7. Wikipedia. SensitiveHighResolutionIonMicroprobe. https://de.wikipedia.org/wiki/Sensitive_High_Resolution_Ion_Microprobe. Accessed: 7. Sept. 2018.
8. Spektrum Akademischer Verlag. U-Pb-Methode. https://www.spektrum.de/lexikon/geowissenschaften/u-pb-methode/17316. Accessed: 11. Sept. 2018.
9. Panstwowy Instytut Geologiczny Warszawa. Pracownia Mikrosondy Jonowej SHRIMP IIe/MC. https://www.pgi.gov.pl/dokumenty-pig-pib-all/foldery-instytutowe/2645-shrimp-folder2/file.html. Accessed: 27 Sept. 2018.

Zircon Sand: An Important Raw Material 5

Zircon is not only a common mineral whose larger crystals make beautiful gemstones (Chap. 2). It is above all a raw material for many technical materials, for example for the zirconium-based oxide ceramics (Chap. 14). What then is the suitable and sufficiently abundant source of zircon? The industrial production of zircon is usually carried out from the zircon sands extracted by mining, which are processed into the so-called zircon concentrates.

5.1 Occurrence of Zircon Sand

Zircon sand is a natural mixture consisting of zirconium oxide and silicon oxide. Zircon itself can also be seen as a combination of the two oxides, which form a common crystal lattice (Chap. 2). In small amounts, zircon sand also contains hafnium, aluminum, iron, and titanium oxides, as well as traces of uranium and thorium. Figure 5.1 shows a collection of zircon sand grains; the dark brown grains are pure zircon.

Zircon sand was first found in 1895 in a secondary deposit called a placer deposit. In geology, a basic distinction is made between magmatic (formed from magma), sedimentary (formed as a result of deposition) and metamorphic (caused by the transformation of other rocks) deposits of minerals. It is also useful to speak of primary and secondary deposits. Primary deposits are those where the minerals are still in the original bedding and have the original bond with the parent rock (Fig. 2.1a in Chap. 2). Primary deposits are of little importance for the profitable mining of zircon.

In secondary deposits, minerals have been transported from the place of their formation and re-sedimented elsewhere. Rivers in particular can transport rocks over long distances. When the water flow subsides, for example at a river mouth, the heavy and weathering-resistant minerals are deposited in front of the ubiquitous and lighter quartz sand, sorted in the process and enriched in certain places. These minerals include corundum, spinel, cassiterite, and diamond, as well as

© Springer-Verlag GmbH Germany, part of Springer Nature 2022
B. Arnold, *Zircon, Zirconium, Zirconia - Similar Names, Different Materials*,
https://doi.org/10.1007/978-3-662-64269-6_5

Fig. 5.1 Zircon sand. (With the kind permission of the Association for the Dissemination of Scientific Knowledge, Vienna)

zirconium-bearing minerals such as zircon and baddeleyite (Chap. 11). Because of their high hardness and chemical stability, they survive weathering and are deposited according to grain size and density. The most important zircon sand deposits are on the coasts of Australia (e.g., on the beaches off Sydney) and in South Africa. One of the Australian zircon sand mines is not far from Perth, where the oldest zircon on earth was discovered (Sect. 4.3 in Chap. 4).

Zircon sand belongs to a series of minerals called heavy mineral sands. In the evaluation of deposits that also carry zircon, their content of titanium minerals (ilmenite, rutile and leucoxene) almost always plays a greater role than their zircon content. Zircon is often considered only as an accessory rather than a primary mineral of value. Zircon sands are usually prepared by electrostatic separation or suitable chemical processes and then made available for further processes as zircon concentrates with a zircon content of about 99.5%. They consist of grains in the size range 0.06–0.3 mm (Fig. 5.1). The first economic extraction of zircon from the zircon sands took place in 1922.

5.2   Application of Zircon Sand

The processed and concentrated zircon sand, i.e. the natural zircon, is used in many different ways. The uses can generally be divided into two groups, depending on how the zircon is further processed: whether it is simply reprocessed or chemically or thermally converted into other materials. An overview of the most important applications of zircon sand is shown in Fig. 5.2.

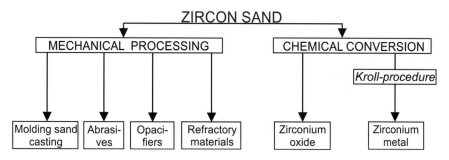

Fig. 5.2 Use of zircon sand. (Based on [3])

On the one hand, zircon sand is used directly in the ground state. Its use is in finely ground form as an opacifier in the ceramic industry. Due to its very high refractive index (Chap. 3), it alters the visual appearance of ceramic parts. A typical application example is the production of opaque porcelain melts for glazes. Zircon opacifiers are also used in sanitary ceramics, glazed bricks and wall, floor and industrial tiles.

Zircon sand has several properties that make it particularly suitable as a refractory material: high melting point, low and regular thermal expansion, good thermal conductivity, chemical resistance and low wettability by molten metals. This results in its possible applications in various fields; it is used, for example, in the lining of blast furnaces.

The refractory properties of zircon, its clean and round grains and its suitability for recycling enable its use as a moulding material in casting technology. As sand (coarser grains) it is used in steel casting and as zircon flour in castings of super alloys and titanium alloys. Here, the fineness of the zircon powder improves the subsequent casting surface and reduces the "burning" of metallic melts.

Furthermore, zircon sand is used in many technical fields as a valuable abrasive.

Secondly, zircon sand is converted into other zirconium-based materials in various chemical and thermal processes. Above all, zircon sand serves as a raw material for the production of powdered zirconium oxide (Sect. 12.1 in Chap. 12), a very important ceramic material today. Only a fairly small proportion of the world's zircon is used to produce metallic zirconium (Sect. 6.2 in Chap. 6). Nevertheless, this use of zircon sand plays an important role, as zirconium is a so-called strategic metal.

Economic aspects (e.g. production, price) of the mining and processing of zircon sand are not dealt with in this book. Reference should be made to the detailed accounts in the relevant literature. In general, zircon sand is not considered a critical mineral, as it is common on almost all continents and its reserves are expected to last for centuries.

Up to now, zircon has been considered a profitable by-product in the mining of ores of the important light metal titanium. Recently, however, interest in zircon, and thus also in zircon sand, as a valuable raw material in its own right has increased significantly. It is, so to speak, the true and irreplaceable source for all zirconium-based materials that we use in practice.

Further Reading

1. Elsner, H. (2006). *Bewertungskriterien für Industrieminerale, Steine und Erden, Teil 12: Schwerminerale. Geologisches Jahrbuch Reihe H, Heft 13* (S. 55–69). Hannover: Bundesanstalt für Geowissenschaften und Rohstoffe und Landesamt für Bergbau, Energie und Geologie.
2. Elsner, H. (2012). *Zirkon—unzureichendes Angebot in der Zukunft? Vortrag beim DERA Rohstoffdialog zur Verfügbarkeit von Zirkon für den Industriestandort Deutschland.* Bundesanstalt für Geowissenschaften und Rohstoffe.
3. Roberts, J. (2015). Zirkonbedarf in der Keramikindustrie und verwandten Märkten. *Keramische Zeitschrift, 67,* 144–147.
4. Schumann, W. (2017). *Edelsteine und Schmucksteine* (S. 62–63). München: BLV Buchverlag.
5. Zirkon—rohwirtschaftliche Steckbriefe. (2013). Bundesanstalt für Geowissenschaften und Rohstoffe, Hannover.

Zirconium: A Hardly Known Metal

6

Having dealt with zircon, we now turn to zirconium, a metal discovered hundreds of years after zircon, but in some ways with its help.

There is more zirconium than copper in the earth's crust. We have known and used copper for thousands of years, zirconium, on the other hand, is basically unknown to this day. Surely this is also related to the fact that it has only a limited and rather very specific application.

Alongside other relatively unknown metals such as indium, gallium, bismuth, tantalum, tellurium and cobalt, zirconium is counted among the so-called "strategic metals". They are important for the development of modern communication technology, electromobility and the energy transition and are so valuable because they are very difficult to replace by other materials.

6.1 Discovery and Naming

In the periodic table of the elements, zirconium (Zr) occupies the 40th place, i.e. its atomic number is 40. It belongs to the IV. Group, the titanium group. In addition to titanium (Ti, atomic number 22) and zirconium, this group also includes hafnium (Hf, atomic number 72) and a radioactive artificial element called rutherfordium (Rf, atomic number 104).

Since hafnium is a twin brother of zirconium and plays an important role in its use, it is described in more detail in Chap. 7. Rutherfordium was first discovered as a transactinoid element in 1964 at the Soviet Nuclear Research Centre in Dubna and named "Kurchatovium" in honour of the Soviet atomic scientist J. V. Kurtschatow. However, the name was rejected in Western countries. In 1969, American researchers identified this element and proposed the name rutherfordium. It was not until 1997 that this element name was agreed upon, and the first one was abandoned.

Titanium is one of the ten most common elements in the earth's crust with a share of approx. 0.4%. Zirconium is much rarer with about 0.02%; however, it is still

© Springer-Verlag GmbH Germany, part of Springer Nature 2022
B. Arnold, *Zircon, Zirconium, Zirconia - Similar Names, Different Materials*,
https://doi.org/10.1007/978-3-662-64269-6_6

found in the group of the twenty most common elements. The metal occurs more frequently than the already mentioned copper or as nickel, cobalt and zinc. Hafnium is rare at 0.0004%, but still more common than better known elements such as tungsten or tantalum. Rutherfordium is produced exclusively in nuclear reactions and only in extremely small quantities.

The discovery of zirconium is attributed to the famous German chemist H. M. Klaproth (1743–1817). However, things were a little different when looked at more closely. In his time, scientists were often concerned with the question of where gemstones got their characteristic color. Nicolas L. Vauguelin, for example, examined rubies and beryls (emeralds) and proved that the coloring agent in both gemstones is chromium. Klaproth was also very interested in gemstones. He had never studied, but acquired his extensive knowledge while working in several pharmacies. In 1810 Klaproth was appointed as the first full professor of chemistry at the newly founded Berlin University. His achievements include the discovery of many chemical elements such as uranium, titanium, cerium, and tellurium (the first representation). Klaproth introduced the balance as a standard analytical instrument. His special preference, however, was mineral analysis. For example, he analyzed the gemstone sapphire and determined that it was aluminum oxide. In his possession was a large collection of minerals, which included nearly 4900 pieces. In 1789, the mineral zircon (more specifically, its yellow-brown variety called hyacinth) from Ceylon (now Sri Lanka) found its way into his laboratory. After analyzing the mineral, however, Klaproth isolated not elemental zirconium but zirconium oxide. He called the highly contaminated oxide "zircon earth". Klaproth gave the name "zirconium", derived from zircon, to the still unknown metal that had to belong to the oxide. Thus the name of a steel-grey metal came from a beautiful coloured gemstone, which in turn was named after its golden colour (Chap. 3).

Zirconium as a chemical element remained unrecognized for a long time and could only be represented in 1824 by J. J. Berzelius (1779–1848) in the form of a black powder. He also proposed the chemical symbol "Zr". He also gave the element symbols to many other elements. Berzelius was a Swedish chemist and physician and is considered the father of modern chemistry. He was ten years old when Klaproth analyzed zircon and formally discovered zirconium. Interestingly, Berzelius was to succeed Klaproth at the University of Berlin in 1817; however, he declined the appointment.

Berzelius succeeded in preparing pure zirconium by reducing completely dry potassium zirconium fluoride (K_2ZrF_6) with potassium. This fluoride was produced by dissolving zirconium oxide in dilute hydrofluoric acid. First, Berzelius heated the fluoride in an iron tube. After treating it with water, drying it, and heating it again (this time with dilute hydrochloric acid), he obtained what he wrote was "a lumpy powder, which was black like coal." A few more steps were necessary until, after a long time, the metallic zirconium was deposited. Later, Antoine Cesar Becquerel succeeded in extracting the metal electrolytically from a zirconium chloride solution in the form of steel-green lamellae.

The correct atomic mass of zirconium could only be determined one hundred years after Berzelius' success in 1924. The reason for this—apart from errors in carrying out the measurements—was the fact, unknown at the time, that zirconium always contains small amounts of hafnium. Without this information, the measurements always resulted in a slightly too high atomic mass.

In conclusion, let us keep in mind that zirconium was discovered in the golden age of chemistry. In the second half of the eighteenth century, several important metals such as nickel (1751), manganese (1774), molybdenum (1778), titanium (1795), chromium (1797), tellurium (1783), tungsten (1783), beryllium (1797) were located in quick succession, including some by analysis of gemstones, in which many scientists were very interested at the time.

6.2 Extraction of Zirconium

The first preparation of zirconium by Berzelius (Sect. 6.1) was based on the reduction of a halide with a metal. Even today, such a reaction forms the basis for the industrial production of zirconium. The extraction of zirconium is complex and similar to that of the homologous titanium.

As already mentioned, zirconium is relatively abundant in the earth's crust, but never elemental. The most important ores are zircon (Chap. 2) and baddeleyite (Chap. 11), which, however, contain only small amounts of the metal. This is why zirconium is often considered rare. On a large scale, we extract it from zircon sand (Chap. 5). Several process steps are necessary for this, which can be summarized in three groups: the conversion of zircon into zirconium oxide, the production of zirconium tetrachloride and the reduction of this chloride.

First, zircon sand is digested in a sodium hydroxide melt and converted into zirconium oxide. Direct reduction of zirconium oxide with carbon (as is done in the extraction of iron in the blast furnace process) is not possible, as the carbides formed in this process are very difficult to separate from the metal. Therefore, the zirconium oxide must be further processed and converted into zirconium carbonitride with coke or graphite in an electric arc furnace. Subsequently, this intermediate product is converted into zirconium tetrachloride by chlorination at 500 °C. According to more recent methods, zirconium tetrachloride can also be obtained directly from zircon concentrates by chlorination in the presence of coke or charcoal. After impurities have been separated, the zirconium tetrachloride is freed from the hafnium chloride, which is also formed, by extraction processes.

The next step is the reduction of the zirconium chloride. This is carried out using the Kroll process in an inert helium atmosphere at approx. 800 °C with magnesium as the reducing agent. The Kroll process was developed by the Luxembourg metallurgist Wilhelm J. Kroll (1889–1973) to obtain technically pure titanium. Since zirconium belongs to the titanium group in the periodic table of the elements, it is chemically similar to titanium. Therefore, the Kroll process can also be used to obtain zirconium. In this process, pure tetrachlorides prepared by the wet chemical route are reduced with molten magnesium in an airtight reaction vessel and under an inert protective gas to form magnesium chloride. Magnesium chloride can be separated by subsequent vacuum distillation at about 900 °C. The product obtained is zirconium in the form of sponge, i.e. a hard, porous mass. Titanium is also obtained as a sponge. In a separate process step (e.g. remelting in a vacuum arc furnace in the absence of oxygen), the zirconium sponge can be converted into compact metal which can be further processed.

If high purity zirconium is required, the thermal decomposition of zirconium tetraiodide according to the Van-Arkel-de-Boer process is used. Zirconium iodide is obtained by heating zirconium with iodine at 200 °C under vacuum. The iodide is then decomposed back into zirconium and iodine on a hot wire at 1300 °C. This process was invented in 1924 by the Dutch chemists A. E. van Arkel and J. H. de Boer for the separation of the purest metals from the gas phase. In this process, the corresponding metal halides, especially the iodides, are evaporated and then split on a wire (e.g. made of tungsten) that is electrically heated to high temperatures. The released halogen can be used to form a new, vaporizable metal halide in colder parts of the apparatus with crude, contaminated metal powder.

The metallic zirconium has many interesting, even extraordinary properties. It therefore has many important applications in technology, whether unalloyed or alloyed with other metals. In Sect. 6.3 we will deal with this topic.

6.3   Properties of Zirconium

In chemically pure form, zirconium is steel grey, soft and ductile. With a density of 6.5 g/cm^3 zirconium belongs to the heavy metals, but it is lighter than steel or nickel alloys. Zirconium is non-magnetic and fully recyclable. Natural zirconium is a mixed element consisting of a total of five isotopes. Due to its position in the periodic table of elements, the metal forms several oxidation states in its compounds, with state IV being the most common and most stable. Oxidation states II and III are present primarily in the solid halides.

As a soft metal, zirconium can be easily machined and polished. It can be easily rolled into sheets and drawn into wires and is also weldable. Its mechanical properties are comparatively moderate: the yield strength is 205 MPa, the tensile strength is 308 MPa and the modulus of elasticity is 99 GPa. The hardness of the metal is increased by impurities. For example, an oxygen content of 0.3% can triple the hardness. Small amounts of hydrogen, carbon or nitrogen make zirconium brittle and difficult to machine. Close control of these elements is therefore necessary so as not to reduce workability.

Zirconium, like the homologous element titanium, is an allotropic metal and occurs in two modifications: the α-(alfa)-zirconium and the β-(beta)-zirconium. At room temperature, the α-zirconium has the hexagonal crystal lattice of densest packing. At the temperature of 870 °C the lattice transformation takes place and the β-(beta)-zirconium with cubic space-centered lattice is formed.

Two properties of zirconium are of particular importance for its technical use: corrosion resistance and permeability to neutrons.

Electrochemically, zirconium with the normal potential of −1.55 V is a base metal. However, it belongs to the interesting group of metals that can passivate. This

Fig. 6.1 Zirconium. (**a**) Just removed from the protective atmosphere; (**b**) After more than one year in air. (© http://images-of-elements.com/zirconium.php CC BY 3.0)

also includes titanium and aluminium as well as nickel and chromium. Passivation means that a metal independently forms a very thin, very dense and well adhering oxide layer, which reliably protects the metal from corrosion. Zirconium also forms a protective zirconium oxide layer very quickly when exposed to air (Fig. 6.1b). This is why it is resistant to almost all acids and bases; only aqua regia and hydrofluoric acid attack zirconium even at room temperature. The corrosion resistance of zirconium even exceeds that of titanium, which is known to be very corrosion resistant. Similar to nickel and chromium, the oxidized surface of zirconium retains the metallic lustre and only becomes a little darker (Fig. 6.1).

Because of its corrosion resistance, zirconium is used in the chemical industry when nickel alloys or stainless steels are no longer sufficient when very aggressive chemicals are used. It is mainly used for special equipment parts such as valves, pumps, pipes and heat exchangers. When used for heat exchangers, its very good thermal conductivity and temperature resistance are additional advantages.

However, the good corrosion resistance can also become a problem, especially when electrochemically machining zirconium. When a positive DC voltage is applied in an aqueous electrolyte—as is common in electrochemical machining— the oxide layer is strengthened in such a way that it forms a barrier to the current. This makes fine electrochemical machining of zirconium more difficult.

The electrochemical formation of the thin oxide layer can be used, as with titanium, for the colouring of zirconium. Depending on the thickness of the layer, different colours are created by double reflection of the incident light and interference in the reflected light.

The good permeability for thermal neutrons is an exceptional property of zirconium. It is characterized by a low effective cross-section or capture cross-section for neutrons or also by a low neutron absorption. The effective cross section is a measure of the probability of the occurrence of a nuclear reaction and indicates the cross section around an atomic nucleus in which a particle, in this case a thermal (i.e. slow) neutron, triggers the reaction. Thermal neutrons are needed for the nuclear fission process in nuclear reactors. There, the fuel rods are kept in tubes made of zirconium, or zirconium alloys, because this metal not only allows the neutrons that drive the nuclear reactor to pass through, but also endures the extreme conditions in the core of a running reactor. This particular application of zirconium is described in more detail in Chap. 9.

The melting point of zirconium is 1855 °C and is higher than that of homologous titanium. At 4410 °C, zirconium has a very high boiling point. These two temperatures indicate a good heat resistance of the metal. This makes it well suited for crucibles, for example, which are incidentally cheaper than crucibles made of platinum.

When it comes to temperatures, zirconium is characterized by another extraordinary property. It burns with the highest temperature for metals of approximately 4650 °C. This produces an impressive and bright flame. In the form of powder, sponge or fine chips, zirconium ignites in air already by friction, impact or electrical discharge. Burning zirconium can only be covered with dry sand (not with water, carbon dioxide or carbon tetrachloride). Because of the danger of ignition, zirconium is stored in argon or methane. Due to the very high temperature, zirconium fires are considered very difficult to extinguish. In principle, most metals are combustible under normal atmospheric conditions, especially alkali and alkaline earth metals. Iron is also combustible in finely divided forms such as steel wool or iron powder. Fine aluminium powder is extremely reactive and ignites explosively by itself on contact with air. Titanium also burns under suitable circumstances. Burns caused by metal fires result in wounds that are difficult to treat and heal poorly.

Because zirconium emits a very bright light when burned, it was used alongside magnesium as a flash powder. This was one of the first practical applications. Unlike magnesium, zirconium burns without smoke, which is why it is used in pyrotechnics and signal lights. Upon impact with metal surfaces, zirconium emits a burst of sparks. This phenomenon is used by the military in some types of ammunition. In film technology, this behaviour is used for non-pyrotechnical impact effects, for example of rifle bullets on metal surfaces.

The electrical conductivity of zirconium, at about 2.4 10^6 S/m, is not as good as that of many other metals; for example, it is only about 4% that of copper. Relative to its poor electrical conductivity, however, zirconium is a relatively good conductor of heat (about 23 W/m K), better than the homologous titanium. Below −272 °C (0.55 K), zirconium becomes superconductive. Zirconium-niobium alloys also exhibit this property and retain it even when strong magnetic fields are applied. Therefore, they are used for superconducting magnets.

Since zirconium reacts with oxygen and nitrogen, it is used as a so-called getter material in incandescent lamps and vacuum systems. At the surface of a getter, gas molecules form a direct chemical bond with the atoms of the getter material or they are held by sorption. In this way, gas molecules are "trapped" and the vacuum is maintained.

Biological functions of zirconium are not known. However, it occurs in small quantities in the human body and is not toxic. Like titanium, zirconium is also biocompatible, which makes it quite interesting for medical technology. However, medical applications are still rare.

The listed and interesting properties make zirconium and its alloys versatile.

Further Reading

1. Sicius, H. (2016). *Titangruppe: Elemente der vierten Nebengruppe* (S. 24–33). Wiesbaden: Springer Fachmedien.
2. Der Brockhaus. (2003). *Naturwissenschaften und Technik* (S. 2238). Heidelberg: Spektrum Akademischer Verlag.
3. Wikipedia. Zirconium. https://de.wikipedia.org/wiki/Zirconium. Accessed: 11. Okt. 2018.

Hafnium: Twin Brother of Zirconium

7

Zirconium has a twin brother—especially geochemically. The brother is called hafnium, is the 72nd chemical element and also belongs to the IV. group of the periodic table of the elements, the titanium group. With zirconium and hafnium we have to do with an extreme case of practically identical ionic radii and charges: The radius of the tetravalent zirconium ion is 86 pm, that of the equally tetravalent hafnium ion is 84 pm. This is due to a phenomenon called lanthanide contraction. As a result, the atomic and ionic radii of the elements from hafnium to rhenium practically do not increase, contrary to expectation. Therefore, both the atoms and the equally charged ions of zirconium and hafnium are almost equal in size. This is the reason for the great chemical similarity of the two elements. It becomes quite clear when the following data are compared: the ionization energy (zirconium: 660 kJ/mol; hafnium: 642 kJ/mol), the electronegativity (zirconium: 1.22; hafnium: 1.23) and the oxidation number, which is four for both elements.

Since many chemical and physical properties of the two elements are almost identical, all zirconium-bearing minerals (zircon, baddeleyite) contain about 1–5% hafnium. In geology, this is referred to as twinning or association of the two elements. Hafnium does not form minerals worthy of mining; although it is not excessively rare (Sect. 6.1 in Chap. 6), it has only been found in three independent minerals. Figure 7.1 shows a zircon crystal from Myanmar that contains much hafnium and also thorite (a rare thorium-uranium silicate).

Besides zirconium and hafnium, yttrium and holmium as well as niobium and tantalum also represent such geochemical twin pairs. The pairs mentioned behave so similarly in many geological processes that they are then described as "controlled by charge and radius".

But still, the twin brothers zirconium and hafnium have quite different properties in some areas. For example, the two metals differ considerably in their density: at 6.5 g/cm^3, zirconium is half as light as hafnium (13.3 g/cm^3). There is another serious difference in the absorption and reflection of neutrons: Zirconium reflects neutrons, whereas hafnium absorbs them. Hafnium has an almost 500-fold larger effective cross-section than zirconium (Sect. 6.3 in Chap. 6) and is impressively described by

© Springer-Verlag GmbH Germany, part of Springer Nature 2022
B. Arnold, *Zircon, Zirconium, Zirconia - Similar Names, Different Materials*,
https://doi.org/10.1007/978-3-662-64269-6_7

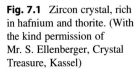

Fig. 7.1 Zircon crystal, rich in hafnium and thorite. (With the kind permission of Mr. S. Ellenberger, Crystal Treasure, Kassel)

experts as a neutron poison. Due to the opposing permeability to neutrons, both metals must be very precisely separated from each other for their use in nuclear technology and must also be highly pure. Suitable separation methods therefore play an important role. Unfortunately, due to their chemical similarity, zirconium and hafnium cannot be separated using standard chemical separation methods. For this purpose, very special methods must be used, e.g. ion exchange or extraction processes, in which the different solubility of zirconium and hafnium compounds in suitable solvents is exploited.

Pure hafnium is a highly lustrous, ductile and fairly soft metal that can be worked. Like its homologous elements zirconium and titanium, hafnium is an allotropic metal. It forms a hexagonal crystal lattice at room temperature, which changes to a cubic one at 1760 °C. Hafnium can passivate, forming a thin, hard and nearly impermeable oxide layer in air. Therefore, it is not attacked by water and alkalis, nor by cold hydrochloric and nitric acids or dilute sulfuric acid. Hot sulfuric acid, aqua regia and especially hydrofluoric acid dissolve hafnium.

Although hafnium is always present as one of its constituents in the mineral zircon, it was traced much later than zirconium. One of the last stable chemical elements to be discovered, it was detected in zircon minerals in 1923 by D. Coster and G. von Hevesy using X-ray spectroscopy. Interestingly, the famous chemist N. Bohr predicted the similarity of the element of atomic number 72 to zirconium a year earlier in his published work on atomic theory. The name hafnium comes from the Latin word "Hafnia" for Copenhagen, because the two researchers there at Niels Bohr's institute interpreted their results correctly.

The extraction of hafnium is very complex, which limits its use. Metallic hafnium was first produced using the Van-Arkel-de-Boer process (Sect. 6.2 in Chap. 6) by depositing hafnium (IV) iodide on a red-hot tungsten wire.

The production volume of hafnium is relatively small, because it has hardly any technical significance. Basically, hafnium is a by-product of zirconium extraction. It is used as a material for control rods in the reactors of nuclear-powered submarines; in metallic alloys (e.g. in tool steels), hafnium can replace other carbide formers.

Further Reading

1. Sicius, H. (2016). *Titangruppe: Elemente der vierten Nebengruppe* (S. 33–39). Wiesbaden: Springer Fachmedien.
2. Wikipedia. Hafnium. https://de.wikipedia.org/wiki/Hafnium. Accessed: 20. Okt. 2018.
3. Spektrum Akademischer Verlag. Hafnium. https://www.spektrum.de/lexikon/chemie/hafnium/3911. Accessed: 8. Jan. 2019.

Zirconium Materials and Their Application

8

Metallic zirconium forms its own alloys as the main element. It also occurs as an alloying element or additive in other materials.

8.1   Zirconium as the Main Alloying Element

In addition to unalloyed (i.e. pure) zirconium, alloys with other metals, especially tin and niobium, are also produced and used. Compared to pure zirconium, they are characterised by improved corrosion resistance as well as higher tensile strength and elongation at break.

Today, semi-finished products made of zirconium materials are mainly used in the chemical and nuclear industries. Which of the materials are suitable for which area depends on its hafnium content. Accordingly, hafnium-containing and hafnium-free zirconium materials are produced and differentiated. The most important zirconium materials and their properties are listed in Table 8.1.

In the case of the unalloyed zirconium materials with the designations Zr700 and Zr702, the mechanical properties are determined by the dissolved gases in the metal, with oxygen playing a particularly important role. In the case of the other alloys, the desired properties are adjusted by selective addition of tin and niobium. The alloys listed in Table 8.1 generally contain small amounts of chromium and iron (together 0.2–0.4%) in addition to the alloying elements. Zirconium materials are basically available in all semi-finished forms. Figure 8.1 shows a tube bundle made of the zirconium material Zr702 for a heat exchanger. Sheets and wires are very good unformable. Unfortunately, the zirconium alloys can become brittle due to the absorption of hydrogen, nitrogen and oxygen, and their corrosion resistance also deteriorates as a result.

Zirconium materials containing hafnium (see Table 8.1) are primarily intended for the chemical industry. Due to the formation of a stable and firmly adhering zirconium oxide layer, they have excellent corrosion resistance, which is highly valued in the chemical industry. The hafnium content plays a subordinate role here.

© Springer-Verlag GmbH Germany, part of Springer Nature 2022   33
B. Arnold, *Zircon, Zirconium, Zirconia - Similar Names, Different Materials*,
https://doi.org/10.1007/978-3-662-64269-6_8

Table 8.1 Properties of some zirconium materials

Specify	Zr700 and Zr702	Zr704	Zr705	Zircalloy4
Alloying elements	Unalloyed Hf max. 4.5%	1.0 ... 2.0% Sn Hf max. 4.5%	2.0 ... 3.0% Nb Hf max. 4.5%	1.2 ... 1.7% Sn Hf max 0.01%
Density in g/cm^3	6.5	6.5	6.5	6.5
E-modulus in GPa	99	95	95	99
Yield strength in MPa	205	250	380	80
Tensile strength in MPa	380	420	580	540
Elongation to fracture in %	16	14	16	28
Thermal conductivity in W/Km	22	17	17.5	28
Application	Chem. industry	Chem. industry	Chem. industry	Nuclear technology

Fig. 8.1 Tube bundle made of zirconium material Zr702 for a heat exchanger. (With the kind permission of the company ASE Apparatebau GmbH, Chemnitz)

Examples of applications for these materials are heat exchangers for acrylic acid plants and reactors for acetic acid plants. Organic acrylic and acetic acids are highly corrosive and extremely aggressive, but at the same time important starting materials for the chemical industry. The demand for acrylic and acetic acids has increased dramatically in recent decades. The production of these two acids requires extremely

Fig. 8.2 Zirconium heat exchanger with explosion-clad tube sheet. (With the kind permission of the company ASE Apparatebau GmbH, Chemnitz)

corrosion-resistant materials for many plant components. No other material except zirconium would be suitable under these conditions. In general, zirconium materials are used in the chemical industry as materials for massive apparatus, heat exchangers, pumps, fittings and pipelines used in the production of phenol, acetic acid, plastics, sulfuric acid and hydrochloric acid media, among others.

Due to its good formability, the low-oxygen zirconium material Zr700 can even be used for explosive cladding. This is a process in which two different types of metal are bonded together over their entire surface using explosives. Usually a less expensive, thicker base material, e.g. steel, is clad with a more valuable, corrosion-resistant material, in this case zirconium. The explosive cladding process creates a metallic bond between the two materials. The tube sheet of the heat exchanger, shown in Fig. 8.2, was explosively clad.

Welding of zirconium materials requires good shielding gas shields and careful monitoring of the shielding gas situation. This is the only way to ensure metallurgically flawless welds with virtually no tarnish. High-purity argon is usually used as shielding gas; the residual oxygen content must always be monitored.

In contrast to the chemical industry, the nuclear industry requires zirconium materials with the lowest possible proportion of hafnium. Its decisive property, as with pure zirconium, is the low capture cross-section (effective cross-section) for thermal neutrons. This means that the neutrons that lead to nuclear fission of uranium in a nuclear power plant are hardly intercepted. These materials bear the name "Zircalloy" (see Table 8.1) and contain tin as the main alloying element and a maximum of 0.010% hafnium. They are used for the cladding tubes of fuel rods (Chap. 9). Their better strength compared with pure zirconium is advantageous. The best-known alloy "Zircalloy2" contains a small amount of nickel (0.07%). The newer alloy "Zircalloy4" lacks nickel, which was dispensed with in order to prevent the absorption of hydrogen as far as possible.

There are zirconium-niobium alloys (e.g. Zr-Nb1, Zr-Nb2.5) which contain even less hafnium than zirconium-niobium alloys (maximum 0.005%). They are also used in nuclear technology. These alloys can be produced in ultra-fine grains and are characterized by biocompatibility. Knee and hip implants have already been made from the alloy Zr-Nb2.5, which have been oxidized beforehand to reduce friction and increase wear resistance.

Zirconium materials have to be melted in an electric arc furnace under vacuum or in an electric beam furnace, which explains the rather high prices. This complex method must also be used for titanium alloys. In the case of hafnium-free zirconium materials, there is also the difficulty of removing hafnium, which further increases the price of the material.

8.2   Zirconium in Other Materials

Zirconium is also found as an alloying element or additive in other materials.

In steel, zirconium exerts a strong metallurgical effect. It has a deoxidizing and desulfurizing effect and also removes nitrogen. In free-cutting steels containing sulphur, zirconium forms sulphides (similar to manganese). Thus the formation of iron sulphides is hindered and the tendency of the steels to red fracture is reduced. Zirconium is a strong carbide former. As a result, it significantly increases the hardness of steel. The corrosion resistance of steel is also improved.

Zirconium increases the service life of heating conductor materials. These materials are used in technology and also in households—wherever heat is needed without long heating. The heating conductor materials find a special application in e-cigarettes.

The addition of zirconium to copper increases its melting temperature without deteriorating the electrical conductivity of copper. A wire drawn from such an alloy is suitable for high voltage equipment.

A magnesium-zinc alloy with small amounts of zirconium remains light but has improved temperature resistance. It also has better strength than an ordinary magnesium alloy and can therefore be used in the manufacture of jet engine components. These turbine jet engines are characterized by high power and thrust with comparatively low masses and sizes.

Biocompatible niobium-zirconium alloys are used to expand the range of metallic implant materials. They contain 1.0–2.5% zirconium and belong to the group of refractory metals. They are characterized by excellent biocompatibility, which is why their medical suitability was already discussed in the 1980s. A necessary strength with good formability is achieved with the help of fine grain hardening. It is important that the modulus of elasticity of these alloys corresponds to that of human bone, which prevents its degeneration on the implant. In addition, these alloys are corrosion resistant.

The group of hard materials includes zirconium nitride (ZrN; zirconium is trivalent here) and zirconium carbide (ZrC). Zirconium nitride is produced by heating zirconium in a nitrogen atmosphere and is used as a refractory material.

Zirconium carbide is produced by reacting zirconium with coke or charcoal and forms very hard crystals that only melt at 3530 °C; this material is used for cutting tools.

Further Reading

1. Wikipedia. Zirconium alloy. https://en.wikipedia.org/wiki/Zirconium_alloy. Accessed: 10. Okt. 2018.
2. ThyssenKrupp AG. Zwei Großprojekte sichern ThyssenKrupp VDM den Eintritt in den Zirkonium-Markt. https://www.thyssenkrupp.com/de/newsroom/pressemeldungen/press-release-48532.html. Accessed: 12. Okt. 2018.
3. Metalcor GmbH. Zirkonium Zr700. http://www.metalcor.de/datenblatt/150/. Accessed: 15. Okt. 2018.
4. Metalcor GmbH. Zirkonium Zr702. http://www.metalcor.de/datenblatt/151/. Accessed: 15. Okt. 2018.
5. Metalcor GmbH. Zirkonium Zr704. http://www.metalcor.de/datenblatt/152/. Accessed: 15. Okt. 2018.
6. Metalcor GmbH. Zirkonium Zr705. http://www.metalcor.de/datenblatt/153/. Accessed: 15. Okt. 2018.
7. Sandvik AB. Sandvik Zr702 Rohre, nahtlos. https://www.materials.sandvik/de/material-center/datenblatter/tube-and-pipe-seamless/sandvik-zr-702/. Accessed: 18. Okt. 2018.
8. ATI Allegeheny Technologies Incorporated. Zirconium Alloys. https://www.atimetals.com/Products/Documents/datasheets/zirconium/alloy/Zr_nuke_waste_disposal_v1.pdf. Accessed: 2. Okt. 2018.
9. ASE Apparatebau GmbH. Einige unsere Erzeugnisse aus und unter Verwendung von Zirkonium. http://www.ase-chemnitz.de/werk_zirkonium_L1.htm. Accessed: 22. Jan. 2019.
10. Rubitschek, F. (2012). Biokompatible ultrafeinkörnige Niob-Zirkonium Legierungen—Integrität unter mechanischer und korrosiver Beanspruchung. http://digital.ub.uni-paderborn.de/hsx/content/titleinfo/516117. Accessed: 26. Jan. 2019.

Zirconium and the Fuel Element

9

Nuclear power plants have a bad reputation, especially in Germany. However, there are many of them around the world and they place high technical demands on plants and materials. Special materials are often needed for some areas. This is the case for fuel elements, for example.

In nuclear power plants, the required heat is generated by nuclear fission. The nuclear fuel usually consists of uranium dioxide, which is used in the form of tablets known as pellets. Nuclear fission takes place in these pellets. The pellets are sintered ("fired" like a ceramic) and then processed very precisely. The term "burning" in the context of nuclear energy ("fuel rod", "fuel element", etc.) is to be understood only in a figurative sense, as this is not combustion, i.e. not oxidation.

The reactor core of a nuclear power plant is made up of several hundred fuel elements. This makes them very important components of nuclear power plants. They are always handled as a whole and they are the first safety barrier in a nuclear reactor.

9.1   Fuel Element and Fuel Rod

A fuel assembly (Fig. 9.1a) is a technically demanding product that operates in a challenging environment. It is exposed to a temperature of more than 300 °C and high pressure. It must withstand severe corrosive stress from fast flowing water. It must have good permeability to thermal neutrons to ensure that nuclear fission proceeds properly. At the same time, it must be resistant to their irradiation. In terms of materials, there is really only one material suitable for this task: Zirconium. As explained in Sect. 6.3 in Chap. 6, zirconium is characterised by very low neutron absorption; it has a small effective cross-section. It also has excellent corrosion resistance and good strength. Neutron irradiation does not cause the metal to become brittle too quickly under normal conditions.

Typical fuel assemblies are about 4 m long and consist of about 250 fuel rods which are bundled together by means of spacers. In most nuclear power plants, the

© Springer-Verlag GmbH Germany, part of Springer Nature 2022
B. Arnold, *Zircon, Zirconium, Zirconia - Similar Names, Different Materials*,
https://doi.org/10.1007/978-3-662-64269-6_9

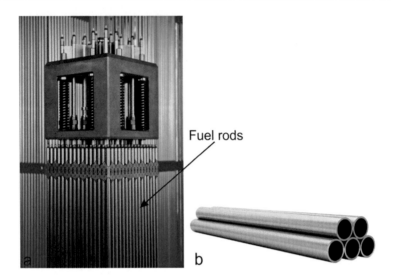

Fig. 9.1 Fuel assembly. (**a**) Model at the Grohnde nuclear power plant, Lower Saxony (© Tobias Kleinschmidt/dpa/picture alliance); (**b**) Zirconium cladding tubes. (With the kind permission of the company Sandvik Materials Technology Deutschland GmbH, Düsseldorf).

fuel rods are surrounded by water as a heat transfer medium. This water therefore transports the usable heat away and cools the fuel rods at the same time. Spacers are used to hold them in precisely defined positions so that the water can flow freely.

A fuel rod is a metallic tube filled with pellets of nuclear fuel. The preferred material for cladding tubes in thermal (e.g. water-cooled) nuclear reactors is the hafnium-free zirconium material zirkalloy (Sect. 8.1 in Chap. 8). The cladding tubes (Fig. 9.1b) have a wall thickness of around 0.6–1.0 mm, depending on the type of fuel element. To achieve good heat transfer in the gap between the nuclear fuel and the tube, it is filled with helium.

The cladding tube of a fuel rod is intended to safely enclose the nuclear fuel, especially to prevent the escape of the highly radioactive fission products. Since these are partly gaseous, the cladding tube must be welded gas-tight. It should remain as tight as possible during operation, despite the high operating temperature, the increasing internal pressure and the possible structural change of the material due to the intensive neutron irradiation.

9.2   Problems with Fuel Elements

The zirconium material zircalloy is used for cladding tubes because of its low neutron capture cross-section, i.e. its high neutron permeability. However, it is problematic that the zirconium can react chemically with water vapour when the fuel rods are strongly overheated, producing explosive hydrogen. Of course, this can only happen if a cooling system has failed and the fuel rods are no longer sufficiently cooled. Significant hydrogen formation can be expected above 450 °C. At these temperatures and in the presence of steam, the zircalloy cladding oxidises. This releases large quantities of hydrogen, which has already led to explosions in the reactor buildings, e.g. in the nuclear disasters at Chernobyl and Fukushima.

Despite its good corrosion resistance, corrosion must also be expected with zircaloy during reactor operation. The thickness of the oxide layer formed here increases steadily over time. It depends on the composition of the material, the cladding tube temperature and the chemical composition of the surrounding cooling water. In addition to damage caused by neutron irradiation, corrosion is one of the processes that limit the service life of the fuel elements in a nuclear reactor to about three to five years. In addition, nuclear fission always converts a proportion of the nuclear fuel into fission products, so that the fuel element can no longer be used effectively to generate energy. It must then be replaced with a new fuel element.

Recently, new fuel rods have been developed that also use zirconium. These new fuel rods are different from the conventional ones. In these fuel rods, the nuclear fuel consists of a uranium-zirconium alloy instead of a sintered uranium oxide, the advantage of which is better heat transport. In this case, the fuel rods are not tubes with pellets, but each is a single piece of metal, fluted and spiral-shaped. This shape allows more water to wet the surface of the fuel rod, dissipating more heat and generating more electricity. At the same time, the larger surface area increases the safety of the reactor core, because nuclear reactions can thus take place at significantly lower temperatures.

Further Reading

1. Humpich, K. (2018). Evolution der Brennstäbe. http://www.nukeklaus.net/2018/03/22/evolution-der-brennstaebe/adminklaus/. Accessed: 21. Sept. 2018.
2. Koelzer, W. (2017). Lexikon zur Kernenergie. https://www.kernenergie.de/kernenergie-wAssets/docs/service/021lexikon.pdf. Accessed: 02. Okt. 2018.
3. Lossau, N. (2011). Welche Atomkraftwerke am sichersten sind. https://www.welt.de/wissenschaft/article12864649/Welche-Atomkraftwerke-am-sichersten-sind.html. Accessed: 30. Sept. 2018.
4. Volkmer, M. (2013). Kernenergie Basiswissen. https://www.kernenergie.de/kernenergie-wAssets/docs/service/018basiswissen.pdf. Accessed: 28. Sept. 2018.
5. Framatome GmbH (ehemals AREVA GmbH). Brennelemente aus Lingen. http://de.areva.com/mini-home/liblocal/docs/PDF-Downloads/ANF-Lingen-Brosch%C3%BCre%20final.pdf. Accessed: 15. Dez. 2018.
6. Paschotta, R. Brennstab. https://www.energie-lexikon.info/brennstab.html. Accessed: 25. Sept. 2018.

Zirconium Oxide: A Versatile Material

10

Of the group of materials dealt with in this book, zirconium oxide represents by far the most important material, especially from a technical point of view. Chemically, it is the most important compound of metallic zirconium (Chap. 6) with predominantly ionic bonds. These bonds are polarized and directional. This explains the great hardness and chemical and thermal resistance of the compound. Its exact chemical name is zirconium dioxide (ZrO_2). The oxide is a polymorphic material. Its lattice transformations play a major role and are described in Chap. 13.

Since zirconium does not form any other oxides, the abbreviated name "zirconium oxide" can also be used. Zirconium oxide is referred to in the technical literature in the abbreviated form as "zircon oxide". For reasons of chemical accuracy and uniform language, only the designation zirconium oxide is used in the book chapters.

The following manifestations, so to speak "faces" of zirconium oxide can be distinguished:

- *Natural zirconium oxide.* Zirconium oxide occurs as a mineral in nature (Chap. 11). The mineral is called baddeleyite or zircon earth. As described in Chap. 2, the German chemist M. Klaproth discovered the oxide in a natural zircon and called it zircon earth at that time.
- *Artificial powdered zirconium oxide.* Today, artificial zirconium oxide (Chap. 12) is the second most important ceramic material after aluminium oxide. It is processed as powder into compact ceramic components or added to paints and other products. The oxide is not synthesized chemically, but is extracted from the mineral raw materials of the earth's crust, above all from zircon sands (Chap. 5). It is often not available in sufficient purity. With the aid of complex washing, cleaning and calcination processes, impurities are separated and pure zirconium oxide powder is obtained.
- *Synthetic zirconium oxide.* Synthetic single crystals with the cubic crystal lattice can be grown from the powdery zirconium oxide. They are called zirconia and,

© Springer-Verlag GmbH Germany, part of Springer Nature 2022
B. Arnold, *Zircon, Zirconium, Zirconia - Similar Names, Different Materials*,
https://doi.org/10.1007/978-3-662-64269-6_10

because of their hardness and brilliance, are mainly used as imitation diamonds in jewellery (Chap. 21).

Regardless of the form in which zirconium oxide occurs and is used, it is characterized by the following properties: It is very hard and wear-resistant, non--magnetic, heat-resistant, insoluble in water and highly resistant to most acids and alkalis.

We will follow this listed classification in the next chapters when describing the properties and possible applications of zirconium oxide.

Natural Zirconium Oxide

11

In nature, zirconium oxide occurs as the mineral baddeleyite, which is quite unknown and rare. It belongs to the group of oxidic minerals, which includes many important ores such as rutile (TiO_2), pyrolusite (MnO_2) and cassiterite (SnO_2). Baddeleyite is chemically homogeneous, but it may contain traces of titanium, hafnium, and iron. At room temperature it has a monoclinic crystal lattice. With a Mohs hardness of 6.5, a density of about 5.8 g/cm^3 and a melting temperature of about 2700 °C, baddeleyite is medium hard, heavy and high melting.

Crystals of the mineral are transparent or translucent and have a high refractive index. An opaque-translucent crystal of baddeleyite from Burma (now Myanmar) is shown in Fig. 11.1. In addition to transparent crystals, baddeleyite can have a brown, green or a mixed colour. Also black baddeleyite crystals are known.

Similar to zircon, baddeleyite is very resistant and can also be used for radiometric age determination. However, baddeleyite does not occur together with zircon in nature (Chap. 2), since it can only form in areas poor in silicon. If silicon is present, then zircon, i.e. a silicate, is formed preferentially.

The name was given in honor of J. Baddeley, who was the first to deal with the mineral from Ceylon (now Sri Lanka). There the mineral was first found in 1892. Baddeley was a geologist, but employed in Ceylon as the intendant of the railway project. Along the way he sent several samples of unknown minerals to the "Museum of Practical Geology" in London, where they were analyzed. Analyses of early samples revealed a new mineral that turned out to be magnesium titanate ($MgTiO_3$), called "Geikielite". Baddeley sent further samples of what he assumed was the same mineral. But analysis of one of these new samples, which was black with a metallic sheen and had a Mohs hardness of 6.5, showed that in this case it was zirconium oxide. The director of the museum suggested that the new mineral be named "baddeleyite" after the person who sent the samples.

Baddeleyite can only be economically extracted from solid deposits, as it occurs in concentrations that are too low in placer deposits. Known deposits of baddeleyite are located in Wyoming in the USA, in the Canadian provinces of Nain and Grenville, in the Vico Volcanic Complex in Italy and in Kodovar in Russia.

© Springer-Verlag GmbH Germany, part of Springer Nature 2022
B. Arnold, *Zircon, Zirconium, Zirconia - Similar Names, Different Materials*,
https://doi.org/10.1007/978-3-662-64269-6_11

Fig. 11.1 Baddeleyite. (With the kind permission of Mr. S. Ellenberger, Crystal Treasure, Kassel)

Due to its high melting temperature, baddeleyite is refractory and is used as a crucible material in the laboratory and in the chemical industry. For example, it is used for the lining of basins in glass melting furnaces. Strictly speaking, the material crystallizing from the melt does not belong to ceramics, but is counted as such.

Metallic zirconium can be obtained from baddeleyite (Sect. 6.2 in Chap. 6). However, baddeleyite is not the most important zirconium ore, but zircon (Chap. 2). The situation here is therefore similar to that of aluminium, which is also not extracted from its natural oxide, corundum, but from another mineral, bauxite.

Further Reading

1. Hülsenberg, D. (2014). *Keramik—Wie ein alter Werkstoff hochmodern wird* (S. 99–100). Berlin: Springer Vieweg.
2. Schorn, S. Mineralienatlas–Fossilienatlas. https://www.mineralienatlas.de/lexikon/index.php/MineralData?mineral=Baddeleyite. Accessed: 5. Aug. 2018.
3. Wikipedia. Baddeleyit. https://de.wikipedia.org/wiki/Baddeleyit. Accessed: 10. Aug. 2018.

12 Artificial Zirconium Oxide

The term "artificial zirconium oxide" refers to a material in powder form intended for sintering. In other words, it is a ceramic material from which components can be manufactured using the sintering technique.

Chemically, zirconium oxide belongs to the group of oxide ceramics. The zirconium and oxygen atoms form both ionic and covalent bonds. The result is a so-called mixed bond with a predominant proportion of ionic bonds (approximately 70%).

Due to its properties, zirconium oxide is counted among the group of high-performance ceramics. In addition to certain properties, high-performance ceramic materials differ from conventional silicate ceramics mainly in the way they are produced, namely by sintering fine powders.

12.1 Powder Production

For technical requirements, it is necessary to produce zirconium oxide using various chemical and thermal processes to ensure the necessary purity and homogeneity of the powder. The most important raw material for zirconium oxide is zircon (zirconium silicate), which can be found as zircon sand (Chap. 5) in heavy metal sands. Baddeleyite (Chap. 11) can also serve as a further raw material source.

In addition to a number of impurities, both minerals also contain 1–3% hafnium (Chap. 7), which is difficult to separate due to its high similarity to zirconium. Often traces of the elements uranium and thorium, which are also present in the minerals, cause low but detectable radioactivity. This is particularly true of baddeleyite, so that the zirconium oxide obtained from it is not suitable for dental technology (Chap. 20), for example.

Because of the required purity, zircon sand is preferred as a raw material for artificial zirconium oxide. The natural zircon sands usually contain magnetite, rutile, ilmenite and other heavy minerals, which have to be removed by magnetic or

B. Arnold, *Zircon, Zirconium, Zirconia - Similar Names, Different Materials*,
https://doi.org/10.1007/978-3-662-64269-6_12

chemical methods. In addition, complex purification steps are required to remove the radioactive substances to a large extent.

How does one obtain the desired zirconium oxide powder from the purified zircon sand, i.e. from the zircon? There are various ways to choose from, which differ mainly in their economic efficiency.

Thus, a thermal decomposition of zircon at about 2000 °C in an electric arc furnace can be applied. During cooling, zirconium oxide separates first, followed by silicon dioxide (SiO_2). The latter is present as solidified quartz glass and must be separated by chemical or mechanical processing. It is therefore better to carry out the decomposition in the presence of carbon. The silicon dioxide is reduced to volatile silicon oxide (SiO), which is subsequently sublimed.

A lower energy input, namely a temperature of only 650 °C, is required for the alkaline decomposition of zircon with caustic soda (sodium hydroxide, NaOH). The zircon is first melted for about two hours with sodium hydroxide as flux. First, sodium zirconate is formed, which is then leached out with water and sulfuric acid. After crystallization, zirconium hydroxide is formed, which is then calcined at 900 °C. After further, partly thermal steps, the desired oxide is formed.

Particularly pure zirconium oxide powders are produced by chlorinating zircon in the presence of carbon at temperatures of 800–1200 °C in a shaft furnace. The resulting volatile zirconium and silicon chlorides can be separated from each other by fractional sublimation. The zirconium chloride is then crystallized with water or calcined to zirconium oxide after purification. Extremely fine powders of high purity can be produced by controlled calcination. In addition, suitable stabilizing agents can be added in this process according to a desired crystal structure of zirconium oxide (Sect. 13.2 in Chap. 13).

The zirconium oxide powder prepared in this way can now be used to manufacture ceramic components by sintering. This powder is one of the so-called non-formable or non-plastic raw materials from which the ceramic masses to be formed are made.

12.2   Sintering of Components

As with all ceramic materials, components made of zirconium oxide are manufactured by sintering. This means that the powder mixtures are homogenized, ground, spray-dried, pressed and then sintered.

In the case of zirconium oxide, the requirement for pure raw materials necessitates a modified preparation compared to conventional ceramics. For the sintering process to run properly, the powders used must have a certain grain size. Plasticizing agents are needed to plasticize the mass, because metal oxides, unlike clays, are non-plastic. Sintering aids must still be added to reduce the residual porosity and thus obtain dense products. After this special preparation, the oxides are processed and sintered by the known methods. In contrast to the production of silicate ceramic materials, this is a dry sintering process, i.e. all reactions take place in the solid state. The possible technological processes are so varied that only a very brief overview can be given here. Please refer to the more detailed descriptions in [1] and [2], for example.

Classic Sintering Technology

The steps typical for the sintering of components are shown in Fig. 12.1.

The shaping of a ceramic mass made of zirconium oxide can be carried out in different ways. The selection of a suitable shaping process is mainly influenced by the shape and geometry of the component to be shaped as well as by the required number of pieces. Uniaxial or isostatic pressing processes can be used for shaping. In uniaxial pressing, the powder is filled into the cavity of the tool formed by the die and lower punch and then pressed into the desired final shape. Isostatic pressing is characterized by the powder being uniformly pressed and compacted in an elastic mold in a pressure vessel containing an incompressible liquid by an equitriaxial pressure.

The quality and homogeneity of the pressed parts depends not only on the pressing process, but also on the powder used and the process parameters. The moulded parts produced by the moulding process are called "green bodies" in ceramics technology. These preforms are dry and may still contain organic auxiliary materials. The term "green machining" is used to describe the machining of these blanks, the aim of which is to produce shapes that are as close to the final contour as possible. Due to the hardness of zirconium oxide, green machining can only be carried out using carbide or diamond tools.

However, the green compacts can also be subjected to heat treatment, i.e. debinding and presintering. During heating, the pressing aid present in the

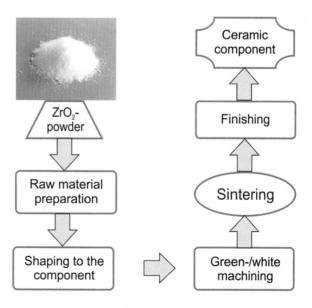

Fig. 12.1 Process steps of the sintering technique

green compacts is thermally decomposed and expelled without residue. The debinding process is followed by a further temperature increase up to the presintering stage, during which contacts are formed between the powder particles by diffusion processes. The pre-sintering (pre-firing) stage is used to adjust important properties of the zirconium oxide blanks such as strength, hardness and shrinkage factor. The resulting parts are referred to as "whites" and their processing accordingly as whitening.

The prepared green and white parts must be sintered. Components made of zirconium oxide are produced exclusively by means of solid phase sintering. The process is based on the diffusion of atoms induced by the supply of thermal energy. Sintering of zirconium oxide requires high temperatures between 1400 and 2000 °C. Initially, the powder particles are held together by adhesive forces. Only through the onset of diffusion are material bridges formed. As diffusion progresses, the powder grains grow together and reduce the pore volume. In the case of zirconium oxide, these processes can be accompanied by component shrinkage of up to 30%.

The process known as the "HIP process" (HIP: hot isostatic pressing) allows the production of a component with virtually 100% dense structure and uniformly fine grain size as well as a high degree of purity. The HIP process is carried out downstream of sintering and takes place at temperatures slightly below the normal sintering temperature under high pressure and in an inert gas atmosphere (argon). The use of the HIP process has been shown to reduce the fracture behaviour of ceramic materials.

The subsequent finishing ensures the required dimensions of the component, which are in line with tolerances. Due to the hardness achieved by sintering, surface machining of zirconium oxide is now only possible with diamond tools, which is associated with high machining costs.

Ceramic Powder Injection Moulding

The modern injection moulding process with CIM technology (CIM: ceramic injection moulding) is particularly suitable for the production of geometrically complex shaped components. In the plastics industry, this process has already been state of the art for many years.

In this process, components made of zirconium oxide with the smallest dimensions and filigree shapes can be manufactured in just one process step. For this purpose, fine ceramic powder is first mixed with a thermoplastic binder material and granulated into an injection-moldable mass. These polymers are mainly based on waxes. The prepared mass can then be processed using methods similar to those used for plastic injection moulding. After the injection molding process, the parts must be debindered again; i.e., the organic binders must be removed before the parts can be sintered. Debinding is carried out quickly at moderate temperatures of around 120 °C in an inert gas atmosphere. The binder is continuously degraded from the outside to the inside. In this way, the gaseous decomposition products of the binder escape through layers that are already porous. The components are then sintered in the sintering furnace at around 1500 °C. During this process, they are subjected to shrinkage of up to During this process, they are subject to shrinkage of up to 30% and are compressed to a maximum. Sintering produces a fine crystalline structure, which gives the material a high fracture toughness and mechanical strength comparable to that of hard metal. Polishing can be used to obtain an extremely smooth surface with low friction values compared to metallic materials.

Ceramic injection molding is not exclusively limited to small, filigree and geometrically complex components. Components with weights of up to 300 *g* and wall thicknesses of 5–6 mm can also be realized with this process.

Further Reading

1. Bargel, H.-J., & Schulze, G. (2000). *Werkstoffkunde* (S. 300–302). Berlin: Springer.
2. Burger, W. (2016). Keramikspritzguss von Hochleistungskeramiken. *Meditronic-Journal, 4,* 16–18.
3. Hülsenberg, D. (2014). *Keramik. Wie ein alter Werkstoff hochmodern wird* (S. 27–67). Berlin: Springer Vieweg.
4. Linsmeier, K.-D. (2010). *Technische Keramik—Werkstoffe für höchste Ansprüche* (S. 15–16). München: Verlag Moderne Industrie Landsberg.
5. Salmang, H., & Scholze, H. (2007). *Keramik* (S. 820–821). Berlin: Springer.
6. Yuheld, Y., Dessy, A., & Nugraha, E. (2016). Processing zirconia trough zirconsand smelting wirh NaOH as a flux. https://jurnal.tekmira.esdm.go.id/index.php/imj/article/view/364/238pdf. Accessed: 18. Okt. 2018.
7. Maxon Motor GmbH. Keramik in der Implantologie. https://www.maxonmotor.de/medias/sys_master/root/8811033002014/Implantate-Dental.pdf?attachment=true. Accessed: 7. Nov. 2018.
8. Coftech GmbH. Zirkonoxid ZrO_2. http://www.coftech.de/produkt/zirkonoxid-zro2/. Accessed: 10. Nov. 2018.

The Crystal World of Zirconium Oxide 13

The crystal world of pure zirconium oxide is diverse. Zirconium oxide is a polymorphic material, just like its basic element zirconium (Sect. 6.3 in Chap. 6). Depending on the temperature, the oxide occurs in three different crystal lattices, which differ in density and thus also in volume.

13.1 Lattice Transformations of Zirconium Oxide

The lattice transformations of zirconium oxide are shown schematically in Fig. 13.1. All transformations take place diffusionless and are reversible, which means that they are martensitic in nature. On the one hand, these lattice transformations cause technical problems, but on the other hand they can also be used in a targeted manner. They make it possible to vary the microstructure and the resulting properties of zirconium oxide-based ceramics over a wide range. The adjustment of the desired microstructure must be carried out on the basis of corresponding and meanwhile well researched state diagrams.

At room temperature, zirconium oxide has a monoclinic lattice. The zirconium oxide occurring in nature, the mineral baddeleyite (Chap. 11), is also monoclinic. The monoclinic lattice is stable up to about 1000 °C. Between 1000 and 1170 °C the transformation into the tetragonal phase takes place. During this transformation of the crystal lattice, the density of the zirconium oxide increases significantly, which in turn means a decrease in volume (see Fig. 13.1). Above about 2370 °C, the tetragonal crystal lattice transforms into the cubic one, which then also remains stable. The density of the cubic zirconium oxide is only slightly greater than that of the tetragonal. Scientists assume that zirconium oxide has generally cubic structure and both the tetragonal and monoclinic phases are distortions of it. The dilatometer curve of zirconia shows hysteresis. While heating, the conversion of the monoclinic phase to the tetragonal occurs at about 1170 °C, whereas during cooling, the reverse conversion occurs at 950 °C.

© Springer-Verlag GmbH Germany, part of Springer Nature 2022
B. Arnold, *Zircon, Zirconium, Zirconia - Similar Names, Different Materials*,
https://doi.org/10.1007/978-3-662-64269-6_13

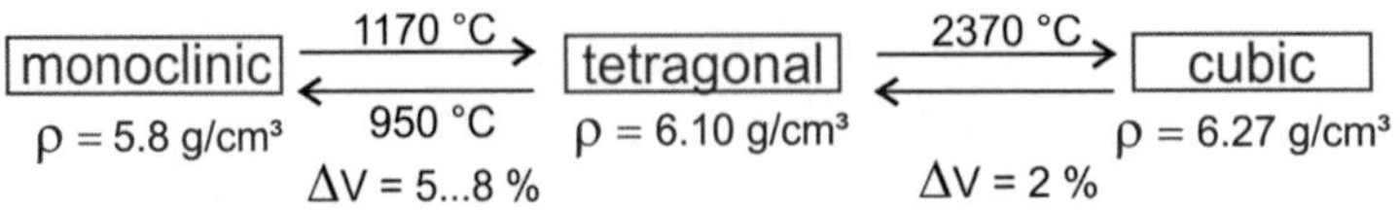

$$\boxed{\text{monoclinic}} \underset{\substack{\xleftarrow{\hspace{1.5cm}} \\ 950\ ^\circ C \\ \Delta V = 5...8\ \%}}{\xrightarrow{1170\ ^\circ C}} \boxed{\text{tetragonal}} \underset{\substack{\xleftarrow{\hspace{1.5cm}} \\ \Delta V = 2\ \%}}{\xrightarrow{2370\ ^\circ C}} \boxed{\text{cubic}}$$

$\rho = 5.8\ \text{g/cm}^3$ $\rho = 6.10\ \text{g/cm}^3$ $\rho = 6.27\ \text{g/cm}^3$

Fig. 13.1 Lattice transformations of zirconium oxide

Of great importance is the transformation of the tetragonal into the monoclinic zirconium oxide, which takes place during cooling. This transformation of the crystal structure is associated with an increase in volume of 5–8%. The resulting stresses lead to the formation of cracks and fissures in the material, and thus it is not possible to manufacture defect-free components. In the worst case, a workpiece can be destroyed.

13.2   Stabilisation of Certain Crystal Lattices

The volume increase during cooling, which is very disturbing for the technical use of zirconium oxide, can be prevented by stabilizing the cubic or tetragonal phase. For this purpose we make use of selective doping. The term comes from the Latin "dotare" and means "to endow". In doping, suitable quantities of foreign atoms are incorporated into the crystal lattice. The prerequisites for this are charge neutrality, mass conservation and conservation of the lattice. The incorporation is more likely if the radii of the incorporated ions do not differ by more than 15% from those of the ions to be replaced and the partners involved have a chemical similarity. The foreign as well as vacant sites created in this way cause the desired changes in properties.

In the case of zirconium oxide, oxides of magnesium, calcium, yttrium and cerium have proven to be stabilizing dopants. Depending on composition and temperature, they form solid solutions with the zirconium oxide. The ions of these elements, especially yttrium, have almost the same ionic radii as the tetravalent zirconium ion and thus fit well into its crystal lattice. The formation of solid solutions causes the shift of the transformation temperatures of the zirconium oxide to lower values. Depending on the amount of oxide added, both the cubic and tetragonal phases can be stabilized. Likewise, partial stabilization is possible in which both phases are present.

When larger amounts (usually from 8 to 15 mol%) of a stabilizing oxide are added, the starting temperature of the lattice transformation is shifted to a very low value. Consequently, the range of existence of the cubic phase expands. This is then maintained down to room temperature. The cubic zirconium oxide is called fully stabilized. The relationships in the lattice can be explained in simplified terms with the aid of Fig. 13.2, which shows the structure of the cubic crystal lattice.

The tetravalent zirconium cations form a face-centered cubic lattice (the Zr^{+4} cation sublattice), into which a cubic-primitive lattice of divalent oxygen anions (the O^{-2} anion sublattice) is inserted. Upon doping and solid solution formation, the added metal cations occupy regular sites of the tetravalent zirconium (Fig. 13.2a).

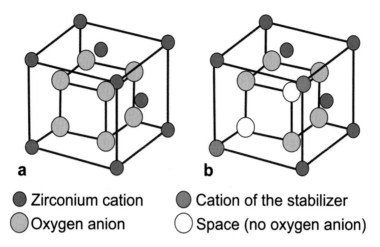

● Zirconium cation   ● Cation of the stabilizer

◐ Oxygen anion   ○ Space (no oxygen anion)

Fig. 13.2 Crystal lattice of cubic zirconium oxide. (**a**) without stabilization, (**b**) with stabilization

However, the oxygen ions belonging to the new cations are not sufficient to fill up the existing anion sites. Since charge neutrality must be maintained at all costs, vacancies are formed in the anion sublattice, i.e. free anion sites (Fig. 13.2b). The vacancies compensate for the residual charge in the lattice and reduce the lattice distortion compared to the non-doped state. Besides, these vacancies cause a very interesting property of the stabilized zirconia: the oxygen-ion conductivity. Depending on the oxygen partial pressure of the environment, oxygen ions can migrate in the crystal lattice and occupy arbitrary vacancies, leading to measurable changes in electrical conductivity. This phenomenon is exploited, for example, in oxygen probes (Chap. 19).

When adding correspondingly small amounts of a stabilizing oxide, the cubic structure cannot form continuously. This process, known as partial stabilization, has proven to be very important in the application of zirconium oxide. With the additional aid of a special and carefully controlled heat treatment, a microstructure can be produced in which only a portion of the grains are in the stable cubic phase. The rest has a tetragonal lattice and is metastable.

For the formation of the metastable state, however, the grains must be very small. The size and also the distribution of these grains can be controlled. Cooling is accompanied by the expected transformation of the tetragonal lattice into the monoclinic one. Due to the associated increase in volume, micro compressive residual stresses build up in the crystal. They inhibit the further transformation, so that the tetragonal phase is partially preserved in the metastable state. Since the residual stresses increase the strength, it is possible to produce higher-strength zirconium oxide materials compared to the monoclinic and non-stabilized variant.

13.3   Transformation Strengthening

Although the increase in volume during the tetragonal/monoclinic transformation of pure zirconium oxide always leads to undesirable crack formation in the ceramic structure, this effect can also be used positively to improve the strength properties of ceramic materials. This phenomenon is referred to as transformation strengthening. The lattice transformation of the tetragonal grains can also be induced by stress. If a sufficiently high mechanical stress is applied, for example in the stress field of a growing microcrack, the tetragonal phase can transform into the monoclinic phase by shear. The transformation that takes place is martensitic and proceeds very rapidly, apparently in an instant. The activation energy is supplied by the very high mechanical stresses applied to the crack tip. The increase in volume caused by the lattice transformation can close, slow down or branch existing cracks, so that the partially stabilized zirconium oxide exhibits a significantly improved fracture toughness (Table 14.2 in Chap. 14). However, this effect only takes place if the critical crack size does not exceed the dimensions of the transformation zone, i.e. the area under stress.

The achievable increase in strength depends on the transformation tendency of the metastable, tetragonal grains. This is influenced by their structure with the respective stabilizer atoms and by their size and quantity. The more tetragonal precipitates are present in the microstructure and the lower the stresses are to trigger the transformation into the monoclinic structure, the greater the fracture toughness. The particle size of the zirconium oxide powder used and its impurities as well as the temperature control during sintering of ceramic components also play an important role (Sect. 12.2 in Chap. 12).

In addition to the high intrinsic strength, the conversion amplification effect also confers an additional safety factor. Colloquially, this effect is often referred to as the "self-healing mechanism" or "airbag effect", because the metastable zirconium oxide particles "inflate" due to the increase in volume and can thus stop crack growth in the ceramic structure.

The transformation enhancement in partially stabilized zirconium oxide proves the potential of lattice transformations from the ground up. A transformation that was initially feared has been ultimately be exploited positively and successfully. The concept of transformation strengthening has led to a serious change in the production of ceramic materials. It was first proposed for zirconium oxide and can also be applied to other ceramics.

Further Reading

1. Aldinger, F., & Weberruß, V. (2010). *Advanced ceramics and future materials—An indroduction to structures, properties, technologies, methods* (S. 123–125). Weinheim: Wiley-VCH.

2. Berek, H. (2012). Weiterentwicklung und Anpassung neuer Methoden der Mikrostrukturanalyse für keramische Systeme mit Phasenumwandlungen. Habilitationsschrift. http://www.qucosa.de/fileadmin/data/qucosa/documents/12793/2013-10-01-Habilitationsschrift%20Berek.pdf/. Accessed: 30. Juli 2018.

3. Decker, S. (2015). Entwicklung der Mikrostruktur und der mechanischen Eigenschaften eines Mg-PSZ-partikelverstärkten TRIP-Matrix-Composits während Spar Plasma Sintering. Dissertation. https://books.google.de/books?id=AmufCwAAQBAJ&pg=PR1&lpg=PR1&dq=decker+sabine+dissertation&source=bl&ots=vec1jnrAQM&sig=nVrPraWPD3MQ1alBRQcQYRxBjUM&hl=de&sa=X&ved=2ahUKEwiKt9XMrNvfAhUGalAKHTDcDKQQ6AEwBXoECAEQAQ#v=onepage&q=decker%20sabine%20dissertation&f=false/. Accessed: 25. Juli 2018.

4. Läpple, V., Drube, B., Wittke, G., & Kammer, C. (2007). *Werkstofftechnik Maschinenbau—Theoretische Grundlagen und praktische Anwendungen* (S. 514–516). Haan-Gruiten: Europa Lehrmittel.

5. Salmang, H., & Scholze, H. (2007). *Keramik* (S. 821–827). Berlin: Springer.

Zirconium Oxide Materials

14

Pure zirconium oxide is almost never used in practice, as the risk of destruction due to lattice transformation is too great (Chap. 13). Therefore, only the stabilized modifications of the oxide are used for ceramic products.

14.1   Classification of Zirconium Oxide Materials

According to the type and quantity of stabilizing elements as well as according to the crystal lattice set up by them, the following material variants of zirconium oxide are distinguished:

- fully stabilized zirconium oxide (CSZ or FSZ, Cubic Stabilized Zirconia) with cubic crystal lattice,
- partially stabilized zirconium oxide (PSZ Partially Stabilized Zirconia) with crystals of cubic and tetragonal phase,
- tetragonal stabilized polycrystalline zirconium oxide (TZP, Tetragonal Zirconia Polycrystal) with tetragonal crystal lattice.

As mentioned in Sect. 13.2 in Chap. 13, the partially stabilised grades of zirconium oxide have a reduced proportion of the added element compared to the fully stabilised form. The composition of zirconium oxide materials and information on their density, porosity and grain size are listed in Table 14.1. Many of the zirconium oxide materials listed above are offered as already doped powders. Ceramic products are manufactured from these powders by sintering (Sect. 12.2 in Chap. 12). However, doping can also be carried out during the sintering of components.

Initially, the increased radioactivity of zirconium oxide ceramics posed a problem in its application. This was due to the uranium and thorium content of the zircon sand raw material (Chap. 5). In the meantime, high-purity raw materials and fine

© Springer-Verlag GmbH Germany, part of Springer Nature 2022
B. Arnold, *Zircon, Zirconium, Zirconia - Similar Names, Different Materials*,
https://doi.org/10.1007/978-3-662-64269-6_14

Table. 14.1 Composition of and information on zirconium oxide materials

Details	CSZ (FSZ)	PSZ	TZP	TZP-A
Components	ZrO_2/Y_2O_3	ZrO_2/MgO	ZrO_2/Y_2O_3	$ZrO_2/Y_2O_3/Al_2O_3$
Composition in %	90/10	96.5/3.5	95/5	95/5/0.25
Density in g/cm^3	5.8	5.7	6.05	6.05
Open porosity in %	0	0	0	0
Grain size in μm	10 Coarse grain	20 ... 50 Coarse grain	0.4 Fine grain	0.35 Fine grain

zirconium oxide powders are available which also meet the requirements for permissible radioactivity.

As mentioned, the corresponding doping is usually already carried out during the production of zirconium oxide powders. This can be done, for example, with the aid of a pulsation reactor, in which a thermal shock treatment takes place under a high heat and mass transfer in an extremely short residence time. A pulsating, high-turbulence hot gas flow is generated in a combustion chamber. The desired reaction, which is specifically terminated by shock cooling (with cold gas), takes place in 0.05–2.0 s. During the treatment, the treatment temperature, treatment time, frequency and amplitude can be specifically adjusted. This influences the particle size, surface properties and phase composition of the material. The difference to other processes lies in particular in the extremely fast heating and cooling rates. The result is extremely fine powders with properties tailored precisely to the application in question. Due to the strong turbulence, each particle experiences exactly the same reaction conditions. In this way, exceptionally homogeneous materials can be produced. Figure 14.1 shows such a homogeneous and fine structure of a zirconium oxide ceramic.

14.2   Fully Stabilized Zirconium Oxide CSZ (FSZ)

In fully stabilized zirconium oxide materials, the concentrations of the doped oxides are selected so high that the cubic phase of the zirconium oxide also exists stably at room temperature. Thus, there is no possibility of transformation strengthening in this material variant and, as a result, the mechanical properties are worse than in the other grades (Table 14.2). For this reason, CSZ ceramics are only used to a limited extent and their characteristic values are difficult to find in the literature or from manufacturers.

The first commercially produced materials of this grade came onto the market in 1928. The first products were crucibles for molten metals, for applications that required such high temperatures that the competing aluminium oxide could no longer be used. The materials used were already stabilized with magnesium oxide (MgO), but only achieved a relatively low sintering density. As an alternative to the stabiliser magnesium oxide, calcium oxide (CaO) was later used with success. Compared to sintered aluminium oxide, however, these fully stabilised zirconium

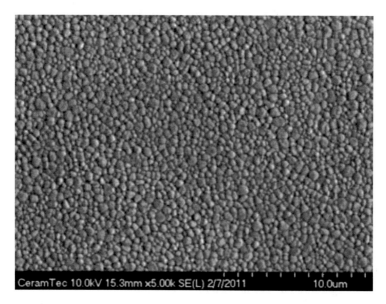

Fig. 14.1 Microstructure of a zirconium oxide ceramic. (With the kind permission of CeramTec GmbH, Plochingen)

Table. 14.2 Mechanical and physical properties of zirconium oxide materials

Property	CSZ(FSZ)	PSZ	TZP	TZP-A
Vickers hardness HV	1200	1200	1300	1300
E-modulus in MPa	150 … 200	200	210	210
4P-Bending strength in MPa	250	500	800 … 1600	1300
Fracture toughness in MPa m$^{1/2}$	k. A	7 … 9	10	10.5
Compressive strength in MPa	2000	2000	2000	2000
Weibull module	k. A	12	20	20
Coefficient of expansion in 10^{-6}/K at 20 °C/1000 °C	k. A	10.5	10	10
Thermal conductivity in W/m-K at 100 °C	k. A	2 … 3	2	2
Specific resistance in Ω cm at 20 °C/1000 °C	k. A.	$10^{10}/10^{3}$	$10^{9}/10^{3}$	$10^{9}/10^{3}$

k.A. No information or information not usual

oxide materials show certain disadvantages, namely a much lower strength and thermal shock resistance. The main advantage is their good oxygen-ion conductivity.

The cubically stabilized zirconium oxide can be grown as a single crystal. These synthetically produced crystals are offered under the name "zirconia" as diamond-like jewellery (Chap. 21). They are characterized by high hardness and high light refraction.

14.3   Partially Stabilized Zirconium Oxide PSZ

The partially stabilized zirconium oxide materials have a cubic matrix structure with tetragonal precipitates (Sect. 13.2 in Chap. 13). Initially, a calcium-stabilized PSZ ceramic (Ca-PSZ) was produced, which was also the first to exhibit transformation enhancement (Sect. 13.3 in Chap. 13). Today, however, magnesium-stabilized zirconium oxide (Mg-PSZ) has gained greater technical importance, as its micro-structure can be more easily adjusted as desired. Here, the optimum content of magnesium oxide is in the range of 8–15 mol%. The tetragonal precipitates are metastable and convertible. With a grain size of about 50 μm, the microstructure is coarsely crystalline. Sintering (Sect. 12.2 in Chap. 12) is to be carried out in a relatively high temperature range in which only the cubic phase exists (temperatures of 1650–1700 °C are required for this), but this results in the formation of the coarsely crystalline microstructure. Complete densification is a prerequisite for successful stress-induced transformation strengthening (Sect. 13.3 in Chap. 13). Sintering is followed by defined cooling or quenching and subsequent annealing in selected temperature ranges. After this, the metastable particles of defined size and homogeneous distribution separate in the cubic grains. Controlled temperature during annealing makes it possible to produce either PSZ materials with maximum strength or materials that are characterized by particularly high thermal shock resistance with lower strength. The chemical resistance of magnesium-stabilized PSZ materials to acids and alkalis—especially when high-purity powders are used— is very good. However, Mg-PSZ decomposes gradually above 900 °C. The unde-sired decomposition of cubic to monoclinic zirconium oxide and to magnesium oxide can be prevented by short holding times during annealing. If PSZ materials are to be used at high temperatures, then this instability must be taken into account.

14.4   Partially Stabilized Polycrystalline Zirconium Oxide TZP

The partially stabilized polycrystalline zirconium oxide materials TZP consist entirely of the tetragonal phase. While high-purity raw materials are not necessarily required for the production of PSZ ceramics, extremely pure and fine-crystalline powders must be used for TZP ceramics (particle size <100 nm). The TZP materials were designed according to the findings of transformation strengthening (Sect. 13.3 in Chap. 13). Their microstructure consists almost exclusively of small, metastable tetragonal grains which, at least in the interior of the material, remain metastable up to room temperature and can transform within the range of action of a possible stress field.

Yttrium oxide (Y_2O_3) and cerium oxide (CeO_2) are the preferred stabilizers, both of which have high solubility in zirconium oxide. The ZrO_2-Y_2O_3 system has been investigated particularly extensively. Of particular technical importance here are the materials containing 3 mol% (or 5 wt %) yttrium oxide, often referred to as "Y-TZP".

The desired microstructure is achieved by specifically adjusting the grain size of the powder and the stabilizer content. In this way, a high-density microstructure can be sintered in which all grains fall below the critical size above which the tetragonal particles are no longer metastable at room temperature. A TZP microstructure should consist of grains with a narrow grain size distribution of less than 0.5 μm, because the tendency to spontaneous lattice transformation and thus to transformation amplification depends on the grain size.

Both the bending strength and the fracture toughness of the Y-TZP ceramics are better than those of the Mg-PSZ ceramics (see Table 14.2). In general, the fracture toughness of the TZP materials is very good and significantly higher than the values of other ceramic materials. The high strength can be further increased by hot isostatic pressing (HIP) (Sect. 12.2 in Chap. 12). After this, the bending strength can reach values of up to 1800 MPa. By comparison, the bending strength of pure zirconium oxide is only 150 MPa. The dense and fine crystalline structure also leads to a low scattering of the strength values, as evidenced by the high Weibull modulus (Table 14.2). The property profile of TZP materials is complemented by their good thermal shock resistance.

It is remarkable that the highest strength is achieved after a grinding treatment of the sintered components. Due to the forces acting during grinding, the tetragonal particles in surface layers are converted into the monoclinic form. The resulting increase in volume leads to compressive stresses on the surface, which further increase the strength. In contrast to other high-performance ceramics, the strength is thus improved by post-treatment, which is always necessary in practice for many components to produce close tolerances.

In a critical temperature range between 200 and 400 °C, the yttrium-stabilized zirconium oxide ceramic can decompose. If the microstructure is unfavourable (defects due to impurities, uneven distribution of the stabilizer, sintering temperatures that are too high), the probability of decomposition becomes greater. In contrast, the Y-TZP ceramic remains stable at higher temperatures.

In addition to yttrium oxide, cerium oxide is also used for the stabilization of TZP ceramics; its usual content is 12 mol%. As with TZP materials, very fine powders are sintered here. Compared with Y-TZP, Ce-TZP ceramics offer higher resistance in a water-containing environment. A small amount of less than 0.5% aluminium oxide can be added to the TZP ceramic. The material is then referred to as TZP-A ceramic and is used in particular for implants in dentistry (Sect. 20.3 in Chap. 20).

The metastable, tetragonal particles of zirconium oxide can also be used in another way. If they are incorporated into a ceramic matrix of another type, for example of aluminium oxide, then the transformation enhancement also leads to a significant improvement in toughness and strength there. A well-known example is zirconium oxide-reinforced aluminium oxide (ZTA) (Sect. 15.2 in Chap. 15).

14.5   Properties of Zirconium Oxide Materials

Table 14.2 lists mechanical and physical properties of the zirconium oxide material variants described in previous sections.

The mechanical properties of zirconium oxide materials are determined by the interaction of microstructure, transformation behavior and defect size. If the transformation of the tetragonal lattice occurs only at a high stress, then the fracture toughness is usually low. If the stress threshold for lattice transformation is low, high fracture toughness is obtained.

The Weibull modulus is a measure of the strength dispersion of a ceramic material. The asymmetric value distribution can be described via the developed statistics. The larger the Weibull modulus, the more homogeneous the structure and the lower the scattering. In the case of ceramic materials, these values are usually determined in the four-point bending test, since tensile tests on specimens made of these materials are costly.

In addition to good mechanical properties, zirconium oxide materials are characterized by very good corrosion resistance, which results from the thermodynamic stability of the zirconium oxide. Interestingly, however, yttrium-stabilized zirconium oxide Y-TZP is often not resistant in a humid atmosphere, e.g. in water vapor. This is referred to as hydrothermal degradation. The degradation process increases from about 150 °C onwards. Presumably, the water vapor dissolves the somewhat larger yttrium ions out of the crystal. As a result, the tetragonal phase becomes unstable and begins to transform into the monoclinic phase at the surface with an increase in volume associated with microcrack formation. Depending on the progress of the transformation, this process is associated with a decrease in strength. The first signs of this can be observed, for example, during the sterilization of surgical instruments at 135 °C. This unfavourable behaviour of Y-TZP can be improved in mixed ceramics by combining it with aluminium oxide (Sect. 15.2 in Chap. 15).

The process of hydrothermal degradation also influences the wear behaviour of the zirconium oxide ceramic. The thermal conductivity of the ceramic materials seems to be of particular importance in this context. If a humid atmosphere and increased stresses also occur during a temperature increase, the transformation of the tetragonal lattice into a monoclinic one can be greatly accelerated. The consequences are increased abrasion and the drop in strength already mentioned. However, general statements regarding a change in properties in a moist or liquid environment are difficult to formulate, since TZP ceramics are produced from different qualities of the starting powder and are processed differently.

14.6 Other Zirconium Oxide Materials

Nanocrystalline Zirconium Oxide Ceramics

Ceramics made of nanocrystalline zirconium oxide are becoming increasingly important as a high-performance material. With a small powder diameter, the diffusion paths that have to be covered by the atoms during sintering (Sect. 12.1 in Chap. 12) are very short. The sintering temperature can therefore be reduced from 1700 to 950 °C when nanocrystalline powders are used. This not only means that the process technology for manufacturing ceramic components is significantly simplified, but also that new material combinations and composite materials based on zirconium oxide are possible. Such materials are used, for example, in components for fuel cells or in sensors. A prerequisite for the use of nanocrystalline ceramics in practice are scalable processes for the production of large quantities of suitable powder. The synthesis of nanocrystalline zirconium oxide is carried out with the aid of chemical vapor phase synthesis (CVS). Optimized processing methods are also important. During sintering of such ceramics, densification and grain growth occur simultaneously. It is therefore difficult to produce dense and concurrent nanocrystalline components.

Particle and Fiber Reinforced Zirconium Oxide Ceramics

Zirconium oxide can be reinforced with metallic particles or with metallic fine fibres. The interaction of a ductile metal introduced into the ceramic matrix with the propagating crack results in an increase in the fracture toughness of the ceramic.

Further Reading

1. Läpple, V., Drube, B., Wittke, G., & Kammer, C. (2007). *Werkstofftechnik Maschinenbau—Theoretische Grundlagen und praktische Anwendungen* (S. 514–516). Haan-Gruiten: Europa Lehrmittel.
2. Liu, T. (1990). Herstellung, Degradation und Ermüdung von umwandlungsverstärkten Y-TZP (A) und Ce-TZP Werkstoffen. Dissertation. https://publikationen.bibliothek.kit.edu/270029412/pdf. Accessed: 17. Juli 2018.
3. Peter, F., Kieback, B., Stephani, G., Andersen, O., & Hermann, M. (2013). Zirkoniumdioxid mit verschiedenen metallischen Partikel- und Feinstfaserverstärkungen. https://www.researchgate.net/publication/326983618_Zirkoniumdioxid_mit_verschiedenen_metallischen_Partikel-und_Feinstfaserverstarkungen. Accessed: 11. Juli 2018.
4. Reckziegel, A. (2015). Eigenschaften und Anwendungen von Hochleistungskeramik aus Zikroniumoxid. https://www.friatec.de/content/friatec/de/Keramik/FRIALIT-DEGUSSIT-Oxidkeramik/Downloads/downloads/FA_Eigenschaften-Zirkonoxid.pdf. Accessed: 25. Juli 2018.
5. Salmang, H., & Scholze, H. (2007). *Keramik* (S. 821–827). Berlin: Springer.

6. Winterer, M. (2004). Nur vom Feinsten—Nanokristalline Keramik. https://www.uni-due.de/ssc/kum/mt_inhalt.php?DID=567. Accessed: 10. Aug. 2018.
7. IBU-tec adavanced materials AG. Der Pulsationsreaktor. Thermisches Verfahren für außergewöhnliche Materialeigenschaften. https://www.ibu-tec.de/anlagen/pulsationsreaktor/. Accessed: 18. Aug. 2018.
8. Metoxit AG. Materialdatenblatt TZP-A. http://www.metoxit.com/assets/Downloads/extern-Datenblatt-TZP-A-DE.PDF. Accessed: 10. Aug. 2018.
9. BCE Specials Ceramics GmbH. Vergleichstabelle. https://www.bce-special-ceramics.de/vergleich/vergleichstabelle.php/. Accessed: 16. Aug. 2018.

Zirconium Oxide Versus Aluminium Oxide 15

Zirconium oxide, along with aluminium oxide, is one of the most widely used oxide high-performance ceramics today. In contrast to silicate ceramics, oxide ceramics are free of silicon oxide; they predominantly have a crystalline structure and are produced exclusively from synthetic starting materials. Oxide ceramics consist of electrically charged anions (always divalent oxygen ions) and multivalent cations (usually metal ions). Their high melting point distinguishes them from other materials composed of oxides, e.g. ferrites and titanates. Understandably, there is no danger of oxidation with oxide ceramics, even at high operating temperatures, since they are already oxidized. They are therefore ideally suited for use in the high-temperature range, for example in firing systems, engines and turbines.

In the field of technical ceramics, the market shares of the two oxide ceramics are different. Aluminium oxide is the absolute market leader with a share of approximately 80%. Zirconium oxide is available on the market with a comparatively small share of approximately 8%, but it is nevertheless of considerable importance for engineering (more on this in Chap. 17). In terms of cost, components made of aluminium oxide are significantly lower than those made of zirconium oxide.

15.1 Comparison of Characteristics

The most important characteristic values of zirconium oxide and aluminium oxide as well as the two mixed ceramics ATZ and ZTA (Sect. 15.2) are listed in Table 15.1.

The comparison of some of the properties of zirconium oxide and aluminium oxide is illustrated in the form of a network diagram in Fig. 15.1.

Using the information in Table 15.1 and the illustration in Fig. 15.1, we can look at and compare the properties of the two oxide ceramics.

Aluminium oxide is characterised by its high hardness, which remains constant over a wide temperature range. This is why aluminium oxide is preferred for cutting tools (indexable inserts). Equally striking is its good thermal conductivity. While aluminium oxide achieves a high value of approximately 30 W/m-K, the thermal

© Springer-Verlag GmbH Germany, part of Springer Nature 2022
B. Arnold, *Zircon, Zirconium, Zirconia - Similar Names, Different Materials*,
https://doi.org/10.1007/978-3-662-64269-6_15

Table 15.1 Characteristic values of aluminium oxide and zirconium oxide as well as ATZ and ZTA mixed ceramics

Property	ZrO$_2$ Y-TZP	Al$_2$O$_3$	ATZ 80 % ZrO$_2$ + 20% Al$_2$O$_3$	ZTA 86% Al$_2$O$_3$ + 14% ZrO$_2$
Density in g/cm^3	6.1	3.9	5.5	4.1
Melting temperature in °C	2680	2050	–	–
Vickers hardness HV	1200	2100	1400	1700
E-modulus in GPa	200	380	220	360
4P Bending strength	1000 ... 1200	600	1400	600
Fracture toughness in MPa m$^{1/2}$	10.5	4.3	5	7
Weibull module	10	10	10	10
Thermal conductivity in W/m-K	3	30	6	25
Coefficient of expansion in 10^{-6}/ K	10	8.5	9	9
Maximum operating temperature in °C	900 ... 1200	1600	1200	1000

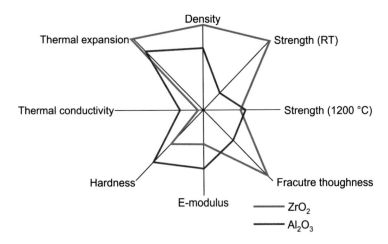

Fig. 15.1 Zirconium oxide and aluminium oxide—comparison of properties

conductivity of zirconium oxide is only one tenth of this. Thus, when zirconium oxide ceramics are used in contact with other materials, there is a risk of local heating due to the frictional heat generated and the poor heat dissipation.

In the case of zirconium oxide, the highest fracture toughness among the oxide ceramics is particularly noteworthy. This fracture resistance—which is significantly greater than that of aluminium oxide—is particularly advantageous in applications where the strength of aluminium oxide ceramics is insufficient. Zirconium oxide is also characterized by high bending strength and thermal expansion, as well as the lowest thermal conductivity among ceramic materials. This property profile makes zirconium oxide ceramics a highly valued ceramic construction material. When dimensioning components made of zirconium oxide, its relatively high density (Table 15.1) must be taken into account accordingly.

The two oxide ceramics are not competing materials, but can support each other, so to speak. This results in mixed ceramics with properties that are better suited to the applications.

15.2 Mixed Ceramics

Mixed ceramics or dispersion ceramics are specifically developed mixtures of different ceramic materials in order to strengthen certain properties and optimize the property profile. Well-known examples of such high-performance ceramics are aluminium oxide-reinforced zirconium oxide (Alumina toughened zirconia, ATZ for short) or zirconium oxide-reinforced aluminium oxide (Zirconia toughened alumina, ZTA for short). The mixture of the two oxides always has a positive effect on the strength values. In the case of mixed ceramics, the effect of the particles embedded in their structures can be compared, figuratively speaking, to the reinforcement of concrete by steel. If the structure of a mixed ceramic is also anisotropic, its strength and fracture toughness can be further increased.

These materials are used for components that must be highly reliable. Mixed ceramics are of particular importance in medical technology. Oxide ceramics have been used successfully in endoprosthetics (knee and hip prostheses) for more than 40 years. Initially, mainly pure aluminium oxide ceramics were used, but these have now been largely replaced by the mixed ceramics ZTA and ATZ. Their wear resistance combined with the bioinert behavior of these materials has largely solved the clinical problem of abrasion-induced aseptic loosening. A good example of such an application of mixed ceramics is hip joint prostheses for orthopaedics (Fig. 15.2).

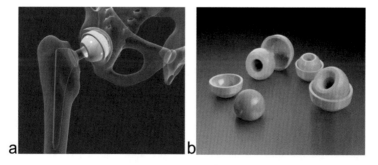

Fig. 15.2 Hip joint prosthesis made of an oxide mixed ceramic. (**a**) General view; (**b**) Condylar heads and cups. (With the kind permission of CeramTec GmbH, Plochingen)

ATZ Mixed Ceramics

The mixed ceramic ATZ is usually based on yttrium-reinforced zirconium oxide Y-TZP (Sect. 14.4 in Chap. 14) and has a higher aluminium oxide content (usually 5 or 20%). Hexagonal platelets of aluminium oxide are embedded in a zirconium oxide matrix. The aim of the mixture is to produce an optimized material that combines the high strength and fracture toughness of zirconium oxide with the hardness of aluminium oxide. Some characteristic values of this ceramic are listed in Table 15.1. The ATZ mixed ceramic has higher hardness and strength than the Y-TZP ceramic, but with somewhat reduced fracture toughness. ATZ mixed ceramics can achieve almost the property profile of hardened tool steel if the chemical composition is suitable. This enables their use for forming tools such as pipe drawing mandrels or for machining tools such as drills.

The addition of more than 10% aluminium oxide improves the hydrothermal properties of zirconium oxide materials. As described in Sect. 14.4 in Chap. 14, the unfavourable behaviour of these materials in a moist environment is a problem. However, mixing the two oxides produces a hydrothermally resistant ceramic that plays an important role in dentistry. There, fully anatomical dental restorations are preferred, but this means direct contact with saliva, i.e. water. Pure zirconium oxide ceramics are not suitable for this purpose because of the risk of subcritical crack propagation. The aluminium oxide content also ensures hot water resistance, which allows components to be sterilised, which is important for medical technology.

One example of a new mixed ceramic is an ATZ material stabilized with cerium oxide (CeO_2). The material was developed for dental applications. The proportion of TZP zirconium oxide stabilized with cerium is 70% and the proportion of aluminum oxide is 30%. A special sintering process of the main components Ce-TZP and aluminium oxide produces an intercrystalline nanostructure. By incorporating Ce-TZP and aluminium oxide crystals with a size of a few nanometres into grains of the other component, an increase in fracture toughness by a factor of 2 was achieved compared to conventional zirconium dioxide ceramics. This good fracture toughness should then not diminish even after years of exposure in humid environments under thermal interaction.

ZTA Mixed Ceramic

Zirconium oxide can be incorporated into an aluminium oxide matrix in the form of metastable, tetragonal particles. In this case, the strength and fracture toughness are also improved compared to pure alumina ceramics. Some characteristic values of ZTA ceramics are listed in Table 15.1.

As described in Sect. 13.3 in Chap. 13, the tetragonal zirconium oxide particles transform to the thermodynamically stable monoclinic modification under stress, which causes the transformation enhancement. Since the coefficient of thermal expansion of the zirconium oxide is higher than that of the aluminium oxide matrix (Table 15.1), radial tensile stresses arise at the boundary between the matrix and the zirconium oxide particles upon cooling from sintering temperature. It is precisely these stresses that can trigger spontaneous lattice transformation. Due to the different thermal expansions of the two oxides, other reinforcing mechanisms are added, but these contribute less to the increase in fracture resistance compared to the transformation enhancement. The combination with the zirconium oxide makes the ZTA ceramic more fracture resistant than the single-phase aluminum oxide ceramic. It retains the good electrical properties of aluminium oxide as well as good thermal conductivity, which changes only insignificantly. This makes the mixed ceramic ideally suited for use as LED substrates, which can also be thin. This compensates for the loss of thermal conductivity and reduces material consumption.

Further Reading

1. Burger, W. (2016). Keramikspritzguss von Hochleistungskeramiken. *Meditronic-Journal, 4,* 16–18.
2. Läpple, V., Drube, B., Wittke, G., & Kammer, C. (2007). *Werkstofftechnik Maschinenbau— Theoretische Grundlagen und praktische Anwendungen* (S. 514–516). Haan-Gruiten: Europa Lehrmittel.
3. BCE Specials Ceramics GmbH. Vergleichstabelle. https://www.bce-special-ceramics.de/vergleich/vergleichstabelle.php/. Accessed: 12. Febr. 2019.
4. CeramTec GmbH. Der universelle Konstruktionswerkstoff. https://www.ceramtec.de/werkstoffe/zirkonoxid/. Accessed: 18. Febr. 2019.
5. Kläger Spritzguss GmH & Co. KG. Hochleistungskeramiken. https://www.klaeger.de/wp-content/uploads/Materialdatenblatt_Technische_Keramik.pdf. Accessed: 5. Febr. 2019.
6. Panasonic. NanoZR pure strength. https://www.phchd.com/global/~/media/dental/global/nanozr/NANO_ZR_Brochure_e.pdf?la=en. Accessed: 7. Febr. 2019.

Ceramic Like Steel

16

In some advertising texts, sintered zirconium oxide, in particular the Y-TZP grade partially stabilized with yttrium, is referred to as "the ceramic steel". This designation is based primarily on the high fracture toughness of this material variant, which is rather untypical for a ceramic material.

In steels, we value the combination of good strength and good toughness. Y-TZP zirconium oxide also exhibits this positive combination of properties. Tensile, compressive and bending strength (Ssee Table 15.1) are good and comparable with the strength of many steels. The only problem is that the values are subject to a scatter that is characteristic of ceramic materials. This scatter is characterized by the characteristic value "Weibull modulus". The partially stabilized zirconium oxide is characterized by a high Weibull modulus (Table 14.2), which indicates a low scatter. The scatter of tensile strength depends on the probability of crack-inducing defects. In the same material, this probability may vary from specimen to specimen; it depends on the manufacturing process and subsequent treatments. Specimen size or volume also affects tensile strength. Compressive strength, on the other hand, is not affected by defects. For this reason, ceramics exhibit much higher compressive strength than tensile strength. For example, Y-TZP zirconium oxide has a compressive strength of 2200 MPa.

When assessing mechanical properties, in addition to the classical characteristic values, information about the behaviour of a material during crack initiation and propagation is also of great importance. The characteristic value used to compare materials with regard to this behavior is the critical stress intensity factor KIc, which is also called fracture toughness or crack toughness. The KIc value characterizes the ability of the material to resist fracture when a crack occurs. It is an experimentally determined material parameter with the unit $MPa\,m^{1/2}$. The higher this parameter is, the safer the material is in the application. The values of the crack toughness of ceramics are significantly lower than those of metals, typically less than $10\,MPa\,m^{1/2}$. If a cracked component is loaded, the KIc value increases with increasing load. Once the externally applied stress reaches the maximum strength, the crack becomes unstable. At this moment, the stress intensity factor reaches a critical value, which is

© Springer-Verlag GmbH Germany, part of Springer Nature 2022
B. Arnold, *Zircon, Zirconium, Zirconia - Similar Names, Different Materials*,
https://doi.org/10.1007/978-3-662-64269-6_16

independent of any further increase in stress. The unstable crack growth usually leads to catastrophic fracture. Therefore, fracture toughness is one of the most important criteria by which ceramics are evaluated.

Due to the transformation enhancement already described (Sect. 13.3 in Chap. 13), partially stabilized zirconium oxide is characterized by an exceptionally high fracture toughness among ceramic materials. In comparison, the KIc value for the most widely used aluminium oxide ceramic is only 4.5 MPa m$^{1/2}$, and unstabilized cubic zirconium oxide has a KIc value of 3.0–3.5 MPa m$^{1/2}$.

This good fracture toughness, combined with good strength, allows us to consider zirconium oxide as the steel among ceramics. The material can be used for the highest requirements in terms of strength, hardness or even toughness. Of course, the fracture toughness of zirconium oxide is far from that of steel. Its KIc values, depending on composition and heat treatment, are between 50 and 90 MPa m$^{1/2}$. However, when compared with aluminium (KIc 15 ... 30 MPa m$^{1/2}$), zirconium oxide performs better. Many plastics, e.g. polymethyl methacrylate PMMA (Plexiglas), also have low fracture toughness values.

The similarity of the zirconium oxide with the steel is also evident in two other characteristic values. Thus, its E-modulus is almost equal to that of steel: the E-modulus of Y-TZP is 205 GPa and that of steel is 210 GPa. The second similar characteristic is the coefficient of expansion (for the zirconium oxide 10×10^{-6}/°C, for steel 11×10^{-6}/°C), which means that the two materials have comparable thermal expansion. These properties play an important role in rolling bearings, for example. During rolling contact in needle or ball bearings, zirconium oxide exhibits a deformation due to its E-modulus that is comparable to that of the rolling bearing steel 100 Cr6, and thus produces the same contact area under load and under the effect of frictional heat.

Zirconium oxide can rightly be called the ceramic steel.

Further Reading

1. Callister, W., & Rethwish, D. (2013). *Materialwissenschaften und Werkstofftechnik—Eine Einführung* (S. 441–442). Weinheim: Wiley-VCH.
2. Läpple, V., Drube, B., Wittke, G., & Kammer, C. (2007). *Werkstofftechnik Maschinenbau— Theoretische Grundlagen und praktische Anwendungen* (S. 514–516). Haan-Gruiten: Europa Lehrmittel.

Zirconium Oxide in Technology

17

Zirconium oxide with its stabilized variants (Chap. 14) is one of the most versatile ceramic materials. One of the first publications on the use of zirconium oxide ceramic for refractory crucibles dates from 1914. Zirconium oxide (with the addition of yttrium oxide) also found an early application as a filament in the Nernst lamp, a new electric incandescent lamp in which its oxygen-ion conductivity was exploited. However, the Nernst lamp was soon superseded by the metal filament lamp.

The still low market shares of zirconium oxide ceramics today (approximately 8%) suggest that it is a niche material. However, this ceramic rather has a key function, which means that it makes many modern products and technologies possible in the first place.

In addition to the properties typical of ceramic materials, such as hardness and wear resistance, as well as corrosion and heat resistance, zirconium oxide is also distinguished by properties that are not found in other ceramic materials and enable its special applications. These include high resistance to the propagation of cracks, i.e. fracture toughness, and oxygen-ion conductivity.

Despite all its positive and interesting properties, zirconium oxide ceramics also have negative aspects. These include, above all, hydrothermal degradation, which has already been described in Sect. 14.5 in Chap. 14.

17.1 Application for Cutting Tools

Thanks to its high hardness and wear resistance, zirconium oxide is ideally suited for cutting tools used in various sectors, e.g. in the paper and packaging industry as well as in the textile and woodworking industries. Various types of cutting tools are also used in medical technology and the automotive industry, as well as in food production. Cutting tools made of zirconium oxide ceramics are particularly interesting when—in addition to good wear resistance and edge strength—corrosion resistance and electrical insulation are required.

© Springer-Verlag GmbH Germany, part of Springer Nature 2022
B. Arnold, *Zircon, Zirconium, Zirconia - Similar Names, Different Materials*,
https://doi.org/10.1007/978-3-662-64269-6_17

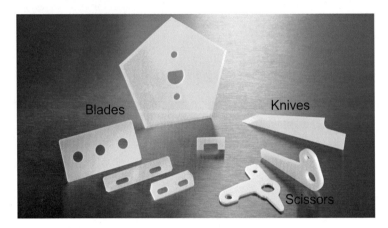

Fig. 17.1 Cutting tools made of zirconium oxide ceramic. (With the kind permission of CeramTec GmbH, Plochingen)

A good example of the use of zirconium oxide as a cutting material are cutting edges for filament and staple fibre yarns in the textile industry. In this field, the yttrium-reinforced zirconium oxide ceramic Y-TZP is used in particular (Sect. 14.4 in Chap. 14). Various cutting tools made of this material can be seen in Fig. 17.1. This zirconium oxide ceramic differs from other ceramic materials in its high cutting edge strength. The corrosion resistance of the ceramic makes it possible to clean the cutting tools with acids or alkalis, thereby significantly extending their service life.

Various components made of zirconium oxide are found in the textile industry. These include thread guides and splicer shears in winding machines. The splicer shears cut out areas that are too thin or too thick, which are determined optically beforehand. In practical tests, the ceramic parts last about four times longer than those made of hard metal. Cutting edges for weaving machines made of zirconium oxide have also already been introduced. While the shears in winding machines only have to manage a few cuts per minute, they cut up to ten times per second in weaving machines.

Since zirconium oxide, in addition to high hardness and wear resistance, also has a certain toughness, it can be used in unusual areas, unlike other ceramic materials. This includes, in particular, the household, where, for example, kitchen knives made of this ceramic are used. Knives also belong to the category of cutting tools. This interesting application is described in more detail in Chap. 18.

17.2   Application for Wear-resistant Components

Another field of application for zirconium oxide ceramics is forming technology, where the tools used are exposed to enormous forces and loads and wear is correspondingly high. A particular advantage of using zirconium oxide is its low coefficient of friction when in contact with metals. This additionally reduces the wear of the tools.

An interesting application example for zirconium oxide are threaded spindles. They are necessary where rotational movements are to be converted into linear ones. Even at high speeds, a ceramic spindle unit runs without problems for a long time and shows no significant wear after millions of cycles. In addition, the use of zirconium oxide virtually eliminates the possibility of galling, i.e. melting of the sliding partners.

Zirconium oxide is being used more and more frequently in the field of pump and engine components. For some time now, a wide variety of pump components have been made from this ceramic, including the pump impeller. This is possible because these components are not usually subjected to dynamic loads. Thus, the toughness of the material plays a subordinate role, although zirconium oxide has a good fracture toughness in contrast to other technical ceramics. Furthermore, the impeller of a pump is not subjected to tensile loads, which is also adapted to the property profile of zirconium oxide ceramics.

17.3   Application in Corrosive Environment

The combination of wear and corrosion resistance allows the use of zirconium oxide in the chemical industry. Components made of this material, for example, ball valves, solid balls of the reversing valves, also meet stringent requirements for functionality and long service life.

A good application example are press dies for the production of tablets from powder in the pharmaceutical industry. However, press dies made of zirconium oxide ceramics are also advantageous in plastics processing and mechanical engineering. Metal-ceramic composite constructions are used in particular, in which the good toughness of the metal is combined with the hardness and corrosion resistance of the ceramic. One example of such a composite design is a press matrix with a core made of zirconium oxide ceramic and a metallic shell. Corrosion and abrasion processes take place within the matrix, for which metals, unlike zirconium oxide ceramics, are not suitable. The metal sheath increases the strength and facilitates assembly.

Cigarette manufacturing machines use a corrosive glue that must be applied to a paper coating in order to seal the cigarette by curing it. The glue is applied via glue rollers, which are thus permanently exposed to the medium. In addition to rollers made of corrosion-resistant steel, glue rollers with running surfaces made of zirconium oxide are now used.

17.4   Application in Case of Thermal Stress

Good heat resistance, low thermal conductivity and high thermal expansion are properties that allow zirconium oxide ceramics to be used for various thermal stresses.

The thermal conductivity of zirconium oxide is one of the lowest among ceramic materials. In combination with its good fracture toughness, the material is a good thermal insulator which, like all ceramic insulators, contributes to uniform temperature distribution.

Due to its low thermal conductivity and high heat resistance, zirconium oxide ceramic is used for the production of kiln furniture. Kiln furniture is equipment of kilns that supports and transports the firing material.

In addition, zirconium oxide has a high coefficient of thermal expansion for ceramic materials, which is similar to that of steel. In frictional connections, this thermal expansion of the zirconium oxide ceramic, which is similar to that of steel, has a positive effect on reducing thermal stresses. This proves to be particularly advantageous in engine construction.

The combination of high thermal expansion and low thermal conductivity leads to interesting tribological properties. Together with its high strength, relatively low modulus of elasticity and high wear resistance, zirconium oxide ceramic is an ideal material for composite steel bearings used at high temperatures. For example, the material combination of zirconium oxide ceramic and steel is found in lubrication-free high-temperature bearings in vehicles. One application example is bearing bushes for exhaust gas valves (Fig. 17.2), which operate at temperatures of approximately 500 °C without any noticeable wear. In the case of the exhaust valve, part of

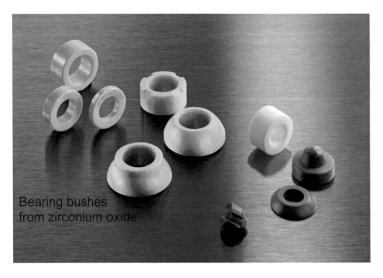

Fig. 17.2 Bearing bushes for exhaust gas flaps made of zirconium oxide ceramic. (With the kind permission of CeramTec GmbH, Plochingen)

Fig. 17.3 Crucible made of zirconium oxide ceramic. (With the kind permission of GTS Gie-ß-Technische-Sonderkeramik, Düsseldorf)

the hot exhaust gas is returned to the intake duct and burned in order to meet the EU exhaust emission standards. Temperatures of more than 450 °C prevail in the vicinity of the engine. Conventional materials fail under these conditions or have a significantly shorter service life.

The property combination of low thermal conductivity and good corrosion resistance plays a decisive role in the use of zirconium oxide for welding shoes. These components are used in the extrusion welding of thick-walled parts made of plastics. Since 2014, welding shoes have been manufactured from zirconium oxide ceramics, which were specifically developed for processing high-melting plastics. Teflon (PTFE) is the main material used for the welding shoe. However, due to its maximum application temperature of 260 °C, it is not possible to process high-temperature plastics with welding shoes made of Teflon. Welding shoes made of zirconium oxide, for example, should enable the welding of high-melting perfluoro-alkoxy copolymers (PFA). The corrosive substances released during the welding of PFA additionally increase the requirements, which are, however, met by zirconium oxide ceramics.

Zirconium oxide is not wetted by most metallic melts. This low reactivity in combination with the heat resistance results in a further field of application. Zirconium oxide ceramics can thus be used to make refractory crucibles, channels, filters, coatings and the like. These ceramic crucibles (Fig. 17.3) have high chemical resistance and, because of the high melting temperature of the zirconium oxide, allow continuous exposure to temperatures in excess of 2000 °C. On the other hand, due to the large thermal expansion of zirconium oxide, they are sensitive to thermal shock and can only be heated up and cooled down slowly.

Furthermore, zirconium oxide ceramic is the best material for welding centering pins thanks to its high heat and wear resistance. It is also used for nozzles in the continuous casting of steel and in the production of metallic powders.

17.5   Application for Tools with Optimized Surfaces

Due to its very good polishability and low friction compared to metallic materials, including steel, zirconium oxide is suitable for tools where the surface finish is important. This is particularly important, for example, in the manufacture of thin wires. The drawing cones and drawing rollers used in this process are exposed to extreme stresses and are subject to high wear. One example of this interesting application of zirconium oxide ceramics is wire drawing tools with optimized surfaces (Fig. 17.4).

For the more or less fine wire variants, raw wire made of copper materials is used. This thin raw wire is produced from a soft cast wire rod, which is drawn over drawing tools with the aid of wire drawing machines until the desired thickness of the wire is achieved. The surfaces of the zirconium oxide drawing tools are polished with diamond tools, as this eliminates surface irregularities from the previous grinding process. The polished surfaces of the ceramic tools have a so-called special cobblestone pattern at the end, which significantly improves the contact with the drawing roller or drawing cone. The smoothly polished surfaces of the zirconium oxide drawing tools also ensure that a lubricant used in the forming process adheres evenly.

Fig. 17.4 Wire drawing tools made of zirconium oxide ceramic. (With the kind permission of CeramTec GmbH, Plochingen)

17.6   Other Areas of Application

Oxygen Sensors

Oxygen-ion conductivity is a very special property of zirconium oxide (Sect. 19.2 in Chap. 19). This property is important where ion movement within a solid material is required, for example in oxygen sensors (Sect. 19.1 in Chap. 19). Zirconium oxide is thus used in the lambda sensors which are widely used today and which are described in more detail in Chap. 19. Such oxygen sensors are used, for example, for continuous measurement of the residual oxygen content in the exhaust gas of industrial furnaces. This special electrical conductivity is absolutely essential for special applications such as in high-temperature fuel cells.

Implants in Medical Technology

Another exceptional property of zirconium oxide is its biocompatibility. Therefore, it is used in medical technology, especially in dental technology. This application is described separately in Chap. 20 due to its special nature and importance. Zirconium oxide is also being used successfully and increasingly in other areas of implant technology—e.g. in endoprosthetics (knee and hip prostheses) (Sect. 15.2 in Chap. 15).

Abrasives and Pigments

Powdered zirconium oxide can be used as an abrasive, similar to aluminium oxide. Due to its hardness and wear resistance, however, it is also frequently used where smooth surfaces need to be protected. For example, zirconium oxide is used to improve the scratch resistance of paints and varnishes, e.g. in automotive topcoats, parquet and furniture varnishes, varnishes for electronic devices, nail varnishes and also in paints for inkjet printers. Because zirconium oxide is white, it is used as a white pigment (similar to titanium dioxide) for porcelain. When mixed with vanadium oxide, it is also used as a yellow pigment.

Insulation Parts in Electrical Engineering

Due to its high electrical resistance (i.e. its low electrical conductivity), zirconium oxide is used in electrical engineering for insulating parts or for high-frequency heating elements. In electronics, ceramic blades made of zirconium oxide prove their worth, as it is not only insulating but also non-magnetic.

Further Reading

1. Kircheldorf, H. (2012). *Menschen und ihre Materialien—von der Steinzeit bis heute* (S. 88). Weinheim: Wiley-VHC.
2. Linsmeier, K.-D. (2010). *Technische Keramik—Werkstoffe für höchste Ansprüche* (S. 11–15, 39, 63, 68, 80). Landsberg: Verlag Moderne Industrie.
3. Oxidkeramik, J. Cardenas GmbH. Hochleistungskeramik Zirkonoxidkeramik CR105 und CR101. http://oxidkeramik.de/oxidkeramik-werkstoffe/zirkonoxidkeramik.aspx. Accessed: 11. Jan. 2019.
4. CeramTec GmbH. Der universelle Konstruktionswerkstoff. https://www.ceramtec.de/werkstoffe/zirkonoxid/. Accessed: 25. Jan. 2019.
5. H. C. Starck Ceramics GmbH. Zirkonoxid-StarCeram Z. https://www.hcstarck-ceramics.de/werkstoffe/zirkonoxid/. Accessed: 25. Jan. 2019.
6. Quitter, D. Schweißschuhe aus Zirkonoxid. https://www.konstruktionspraxis.vogel.de/schweissschuhe-aus-zirkonoxid-a-451451/. Accessed: 14. Jan. 2019.

Zirconium Oxide Kitchen Utensils

18

At the beginning of their technical use, ceramic materials were considered useless for kitchen products such as knives or scissors. However, the development of stabilized grades of zirconium oxide ceramics (Chap. 14) with improved properties has greatly changed the situation. Today, many kitchen utensils made of this ceramic can be found in the kitchen.

Probably the best known example are ceramic knives. They are significantly harder and sharper than knives made of steel and offer high cutting quality. They remain sharp for a particularly long time and can be used for a correspondingly long time. Tedious re-sharpening as with steel knives is rarely necessary. However, diamond tools, which are harder than the zirconium oxide ceramic, must be used for sharpening.

With a ceramic knife made of zirconium oxide, very straight and fine cuts are possible. In addition, they are very light and so you can work with them without fatigue. Another advantage of zirconium oxide is its corrosion resistance. This means that the knives and also other ceramic kitchen utensils are insensitive to oils, acids, juices and salts. When in contact with food, the ceramic does not emit metal ions or other contaminants. It is pure and germ-free and neither the smell, taste nor appearance of the food is changed. The heat and wear resistance of zirconium oxide are also advantageous in kitchen work.

The well-known Japanese company Kyocera produces ceramic knives from two of its own types of zirconium oxide ceramic (white and black) (Fig. 18.1).

The raw materials come from the company's own mine in Australia and are ground particularly fine. These nanoparticles are the basic prerequisite for the sharpness and breaking strength of the knives. The zirconium oxide powder is brought into shape with a high pressure and then sintered at over 1400 °C. The black ceramic is even processed at a very high pressure and sintered a second time at over 1500 °C after a first sintering cycle. In this way, a particularly high-quality knife can be produced. The first grinding of the blanks takes place in a drum with silicon carbide. Then the blades are ground manually over the entire surface in several steps on diamond wheels. In contrast, most steel knives are sharpened by machine.

© Springer-Verlag GmbH Germany, part of Springer Nature 2022
B. Arnold, *Zircon, Zirconium, Zirconia - Similar Names, Different Materials*,
https://doi.org/10.1007/978-3-662-64269-6_18

Fig. 18.1 Ceramic knife in two colours. (With the kind permission of Kyocera Feinceramics GmbH, Neuss)

Since the majority of kitchen knives are steel, we thoughtlessly transfer their properties to all knives. But with ceramics, this can lead to surprises. The downside of hard ceramic blades is their brittleness, which is much greater than steel knives. They chip and break easily. If they hit something hard or fall to the ground, the ceramic knife may be damaged.

In addition to ceramic knives, other tools suitable for the kitchen are made from zirconium oxide. These include fruit and vegetable peelers, slicers and graters as well as kitchen scissors, which make it very easy to work evenly thanks to their sharp ceramic blades. They are all additionally germ-free and thus meet high hygienic requirements. Cleaning scrapers have two sides, a ceramic side for hard surfaces and a softer side made of plastic for softer surfaces.

Various mills with ceramic grinders that are suitable for salt, pepper and other spices as well as for coffee are widely used and popular. With the help of numerous stages, the material to be ground can be reduced to almost any size. In contrast to the steel grinder, the grist is grinding in the ceramic grinder and not cut.

Although ceramic grinders are less sensitive to moisture than steel grinders, the small teeth of the grinder can break if used incorrectly. It is therefore particularly important that mills with ceramic grinders are never operated empty. Direct rubbing of the fine ceramic teeth against each other can cause damage. You should pay particular attention to this with brand new mills and also with mills whose grinding stock is already running low.

In advertising for ceramic kitchen gadgets, the slogan "Back to the Stone Age" is used from time to time. Interestingly enough, it is quite accurate. In terms of materials, stones and advanced ceramics can be grouped together under the generic term "non-metallic inorganic materials".

Further Reading

1. KYOCERA Fineceramics GmbH Keramikmesser. https://germany.kyocera.com/index/products/kitchen_products/ceramic_knifes.html. Accessed: 27. Nov. 2018.
2. Ronicke Ph. Pfeffermühle mit Keramikmahlwerk. https://www.pfeffermuehle-test.de/mahlwerk/keramikmahlwerk/. Accessed: 28. Nov. 2018.

Zirconium Oxide and the Lambda Sensor 19

A so-called lambda sensor can be found in almost every car today. And perhaps you have already asked yourself what a lambda sensor actually is, what it does in the car and how it really works?

19.1 Task of the Lambda Sensor

The lambda sensor is an important measuring device for the exhaust gas purification of vehicles with petrol engines. Three-way catalytic converters have been mandatory in Germany since 1993. They reduce the proportion of environmentally harmful and partly toxic pollutants in the exhaust gas. Through complex chemical reactions, the vehicle catalytic converter converts carbon monoxide, nitrogen oxides and unburned hydrocarbons into carbon dioxide, nitrogen and water. At operating temperature and with an optimum oxygen content in the exhaust gas, catalytic converters today achieve an efficiency of practically one hundred percent. The task of the lambda sensor is to provide the required information for these chemical processes. In other words, the lambda sensor and the catalytic converter ensure clean air. Figure 19.1 shows an exhaust gas cleaning system with three-way catalytic converter and lambda sensor for a Mercedes 300E.

The lambda sensor is actually an oxygen sensor. It measures the oxygen content of the exhaust gas and transmits the value in the form of an electrical signal to the engine control unit. The measured value from the lambda sensor enables the control unit to regulate the injection quantity in such a way that the best possible composition of the combustion mixture is ensured. This creates ideal conditions for exhaust gas treatment in the catalytic converter. Emissions of pollutants always occur when the oxygen content deviates from the optimum value. That is why its determination is so important.

Where does the name of this probe come from? The lambda sensor is so named because it measures an important variable, the combustion air ratio λ (Greek: lambda). In combustion theory, this describes the mass ratio of air and fuel in a

© Springer-Verlag GmbH Germany, part of Springer Nature 2022
B. Arnold, *Zircon, Zirconium, Zirconia - Similar Names, Different Materials*,
https://doi.org/10.1007/978-3-662-64269-6_19

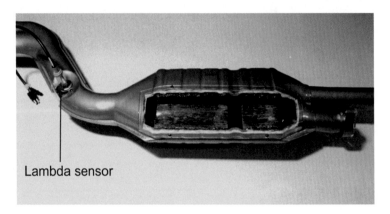

Fig. 19.1 Three-way catalytic converter with lambda sensor. (©Harry Melchert/dpa/picture alliance)

combustion process (air-fuel ratio or the "air number"). For the three-way catalytic converter to function properly, a control circuit is used to keep this ratio within a narrow range, the so-called lambda window.

In order to measure the oxygen content, the lambda sensor must contain an element that can conduct oxygen ions well. And this is where zirconium oxide comes into play. A conductive membrane made of this material is found above all in the widely used jump sensors. How zirconium oxide becomes a conductor of oxygen ions has already been mentioned in Sect. 13.2 in Chap. 13 in the description of crystal lattices. Let us briefly revisit this topic.

19.2   Oxygen-ion Conductivity of Zirconium Oxide

The specific electrical resistance of zirconium oxide is extremely high compared to that of metals, making it an insulator. This is due to the fact that the electrical charges in the mainly heteropolar-bonded material are fixed to ions. These in turn are firmly localized in the crystal lattice. However, the electrical resistance decreases with increasing temperature. Diffusion allows extremely low ionic conductivity and therefore a ceramic insulator can be an electrical conductor. The conductivity is determined by the diffusion coefficients of the ions. These are characteristic of the ions but also depend on defects in the crystal lattice. The last effect can be specifically exploited. One builds oxygen vacancies into the crystal lattice. The diffusion of the oxygen anions across these vacancies is associated with a charge transport that is significantly higher than in an insulator.

How are oxygen vacancies created in zirconium oxide? Similar to the improvement of strength, they are achieved by doping (Sect. 13.2 in Chap. 13). New solid solutions are formed as a result of doping processes. Let us take doping with yttrium oxide (Y_2O_3) as an example. When this oxide is incorporated into the crystal lattice of the zirconium oxide, the trivalent yttrium ions (Y^{3+}) occupy the lattice sites of

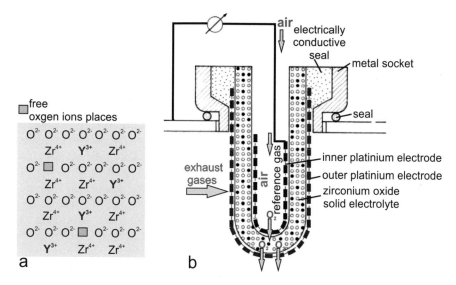

Fig. 19.2 Lambda sensor. (**a**) Charge balance in yttrium-stabilized zirconium oxide (schematic); (**b**) operating principle. (© Martin Olson CC BY 3.0—slightly modified)

tetravalent zirconium ions (Zr^{4+}). However, an yttrium ion requires one less electron than a zirconium ion to balance its positive charge. Without a change, a negatively charged oxygen ion sublattice would result. But this is not possible. When cations are exchanged, just enough vacancies (i.e. free anion sites) are formed in the oxygen ion sublattice for a complete charge balance to occur. In this process, the oxygen ions can "choose" the free lattice sites, so to speak, and become significantly more mobile. This results in the desired charge transport. This charge balance through free anion sites is shown schematically in Fig. 19.2a.

For other gases (atoms and ions) as well as electrons, the doped zirconium oxide is a non-conductor. It is therefore a permeable membrane only for oxygen ions. The highest electrical conductivity is achieved with fully stabilised zirconium oxide (Sect. 14.2 in Chap. 14). However, when measuring with the lambda probe, relatively small voltage differences are measured, for which even a lower electrical conductivity is sufficient. Partially stabilised materials are therefore almost always used, for example the Y-TZP grade (Sect. 14.4 in Chap. 14), which have better mechanical properties than fully stabilised materials.

19.3 Operating Principle of the Lambda Sensor

A directed diffusion of oxygen ions can be triggered by a built-up difference of the oxygen partial pressure. One can also exploit an already existing pressure difference. This possibility exists in the manufacture of sensors for exhaust gas cleaning. The reference pressure results from the oxygen content of the ambient air; the other pressure is present in the exhaust gases. The oxygen ions migrate in this gradient of the partial pressure. A low DC voltage is produced, which depends on the pressure gradient. This is sufficient to determine the oxygen content in exhaust gases and to use its value to control combustion in gasoline engines and for catalytic afterburning. And this brings us back to the lambda sensor. Its operating principle is shown in Fig. 19.2b.

The actual measuring cell consists of two porous electrodes made of a precious metal (usually platinum) and the ion conductor made of the partially stabilised zirconium oxide. Since the ion conductivity increases with rising temperature, the lambda sensor only reaches a favourable operating temperature between 400 and 1000 °C.

In technical lambda sensors, probe on the inside of the sensor can measure the difference in concentration of oxygen between exhaust gas and reference air. The reference air is normally the ambient air (see Fig. 19.2b), which, depending on the design of the sensor, is also supplied separately in order to avoid contamination and incorrect measurement results due to carbon oxides, water, oil or fuel vapours. There are also probes that do not require additional air and instead have an oxygen-free reference inside the sensor as a reference point.

Like any component in a car, a lambda sensor can also break down. Too high temperatures are often the cause of the defect. Chemical stress or mechanical stress, e.g. vibrations of the vehicle, can also cause damage. As a result of the malfunction of the lambda sensor, a car can stop unexpectedly, which once happened to the author of the book.

Various technical designs of lambda sensors for specific car models can be found on the internet, for example. Descriptions of the structure of lambda sensors often refer to a "zircon element". However, the truth is that the material used is not zircon (i.e. zirconium silicate), but zirconium oxide.

Further Reading

1. Hülsenberg, D. (2014). *Keramik—Wie ein alter Werkstoff hochmodern wird* (S. 124–128). Berlin: Springer.
2. Verein Freier Ersatzteilemarkt e.V. Lambdasonde. https://www.mein-autolexikon.de/abgasanlage/lambdasonde.html. Accessed: 5. Sept. 2018.

Zirconium Oxide in Dentistry

20

Zirconium oxide is best known for its medical application in dental technology. Due to its high strength and good fracture toughness, as well as its biocompatibility and especially its colour adaptability, it has steadily gained importance in the fabrication of dental restorations. Zirconium oxide ceramics are used in orthodontics for brackets and in dental prosthetics as a material for root posts, crowns and bridges. In surgery, it can be used as a base material for abutments and implants.

Zirconium oxide-based mixed ceramics ATZ (Sect. 15.2 in Chap. 15) are also used in medical technology—for example as a material for manufacturing hearing and finger endoprostheses.

Zirconium oxide is by no means a completely new material in medical technology. It has already been investigated as an implant material in the field of hip joint prosthetics and its favourable long-term behaviour has been proven.

20.1 Requirements for Dental Crowns and Dental Bridges

Applications of zirconium oxide ceramics in dentistry range from conservative restorations (tooth-preserving care) to crowns and bridges to implant prosthetics. In principle, the zirconium oxide used must meet the requirements specified for a surgical implant material in the corresponding standard [5]. Among other things, this standard sets strict requirements for the radioactivity of raw materials. For example, it must not exceed the limit value of 200 Bq/kg, which is considered safe for materials in the human body. From tests of the radioactivity of raw material batches of zirconium oxide, it is known that these values are complied with in practice or are even undercut in some cases.

The advantages of zirconium oxide as a tooth replacement material lie in its excellent biocompatibility and its tooth-like colour, which meets aesthetic requirements. The good biocompatibility of zirconium oxide is based, among other things, on its low solubility in various media. In terms of aesthetics, only the opacity (i.e. lack of light transmission) of the zirconium oxide commonly used in

B. Arnold, *Zircon, Zirconium, Zirconia - Similar Names, Different Materials*, https://doi.org/10.1007/978-3-662-64269-6_20

technical practice has a detrimental effect. For this reason, the oxide was initially used mostly for posterior teeth without subsequent ceramic veneering. With the development of new types of zirconium oxide, which are even more similar to natural teeth in their optical properties due to their translucency (partial light transmission), the range of applications has been extended. The ceramic dental material made of zirconium oxide is available in three basic shades (light, medium, intensive). In the case of a monolithic crown or bridge, a more precise shade match can be made with stains. If only the crown or bridge framework is made of zirconium oxide, which is then customized with a fired-on ceramic veneer, the most sophisticated aesthetic results can be achieved. By changing the shades, however, it is now possible to match zirconium oxide crowns and bridges to the shade nuances of the remaining dentition in such a way that subsequent veneering can be dispensed with.

However, the concessions made to aesthetics inevitably affect the material properties. For example, translucent zirconium oxide grades have a lower fracture toughness, which somewhat limits their range of applications compared to conventional zirconium oxide.

The low thermal conductivity of zirconium oxide (Table 14.2 in Chap. 14) corresponds to that of natural tooth substance. This is advantageous because it results in less thermal irritation. The wearing comfort is very high, as a dental crown made of zirconium oxide does not transmit heat or cold to the dental nerves to such an extent.

Zirconium oxide—unlike a large proportion of other ceramic materials—can also be cemented with conventional cements based on zinc phosphate or glass ionomer. This is another advantage of this material.

The mechanical properties such as hardness and bending strength as well as modulus of elasticity (Table 14.2 in Chap. 14) also speak for the suitability of zirconium oxide in dental technology. As described in (Sect. 13.3 in Chap. 13), the partially stabilized tetragonal phase of the oxide allows a special strengthening mechanism to be realized, the so-called transformation strengthening. Thus, tensile stresses occurring at crack tips in the material can be relieved by a transformation of the tetragonal phase into the monoclinic phase. Due to the approximately 5% higher volume of the monoclinic phase, the crack is "clamped" and subcritical crack growth is prevented.

However, the high hardness of the zirconium oxide, which is higher than that of the natural tooth substance, may prove to be disadvantageous, as the teeth of the opposing jaw may be damaged in case of pronounced grinding and/or pressing. In addition, the risk of so-called chipping, i.e. shearing off of the veneer from the basic framework of a crown or bridge, is significantly higher. The high wear resistance of zirconium oxide also has a certain disadvantage. Under certain circumstances, a monolithic crown made of zirconium oxide can lead to severe chipping of teeth in the opposing jaw.

By the way: In dental technology, we often speak of a "zircon" crown or a "zircon oxide" crown. However, the correct term "zirconium oxide" crown is usually avoided. Why is that? According to dentists, the word "zirconium" has negative connotations because it is mistakenly associated with a metal, i.e. a metallic crown or bridge. Unfortunately, the second component of the term, namely the word "oxide", is ignored. And that oxides are not metals is unfortunately hardly known.

20.2 Fabrication of Zirconium Oxide Ceramic Dentures

The properties of zirconium oxide do not allow the use of the classical sintering technique, which is based on the direct processing of powders (Sect. 12.2 in Chap. 12), in a dental laboratory. For the fabrication of dental restorations, the zirconium oxide ceramic is used in one of the possible prefabricated states shown in Fig. 20.1.

The typical yttrium-stabilised zirconium oxide material Y-TZP (Sect. 14.4 in Chap. 14) is nowadays predominantly machined as green compact or white compact in a pre-sintered, open-pored and chalk-like state. Due to a relatively low material density, these blanks can still be easily machined. The crown or bridge frameworks to be fabricated are machined from the blanks using carbide burs or diamond grinding tools. The shrinkage factor of the ceramic material must be taken into account, i.e. the workpieces are prefabricated approx. 25% larger. The prepared dental prosthesis parts are then finished by sintering. In this way, the final material properties are achieved with the previously calculated shrinkage and the pore reduction that takes place.

In recent years, the development and increasing use of CAD and CAM methods in dentistry have changed prosthetic technology (CAD: computer aided design; CAM: computer aided manufacturing). The application of these modern methods involves the use of industrially produced blanks, so-called dental blanks (Fig. 20.2).

Such dental blanks are characterized by good machinability, constant shrinkage and high edge stability. Thanks to special shaping methods and precise process control during pre-compaction and pre-sintering, they also have the right density and hardness so that they can be machined safely and highly efficiently in a short time. The dentures are fabricated in automated milling units according to computer-aided design and can be made both before and after the final sintering process of the blanks.

However, the majority of laboratory production of prosthetic workpieces is still based on the traditional processing of green and white parts. In contrast to industrial

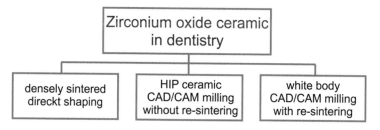

Fig. 20.1 Zirconium oxide states for the fabrication of dentures. (Based on [2])

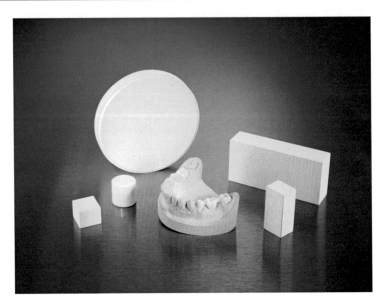

Fig. 20.2 Dental blanks. (With the kind permission of CeramTech GmbH, Plochingen)

production, dental laboratories always produce unique and never series-produced identical products. However, computer-controlled planning, fabrication and processing of dental crowns and bridges generally have advantages, so CAD and CAM methods will certainly be used more in the future.

20.3   Implants

Zirconium oxide has also proven suitable as a material for endosseous implants, i.e. implants placed inside the jawbone. This application is made possible by the aforementioned combination of properties of this material. Interestingly, alumina ceramics have not been able to establish themselves as an implant material in the dental field. Ceramic implants are usually made of an yttrium-stabilized zirconium oxide ceramic TZP-A (Sect. 14.4 in Chap. 14), which contains a small amount of aluminum oxide. The required material properties can be optimized by the special HIP process (Sect. 12.2 in Chap. 12). In this manufacturing step, the ceramic is recompacted for several days at high pressure after sintering in a tunnel kiln.

Zirconium oxide implants of the latest generation show several advantages over the established titanium implants. White zirconium oxide implants are much closer to the natural tooth colour than titanium implants, which is an aesthetic advantage especially for restorations with thin gums. Healing into the bone is equivalent to that of titanium implants, but significantly longer healing times must be taken into account. In the contact area with the gumline, zirconium oxide implants are even better, as the accumulation of bacteria is lower.

However, zirconium oxide implants also have some disadvantages. These include higher costs and still little scientific knowledge as well as a lack of long-term experience. Aging properties of zirconium oxide are still viewed critically. The optimal surface properties of zirconium oxide implants for osseointegration (integration of an implant into the bone) have also not yet been fully clarified. For this purpose, the implants must have a special surface structure that corresponds to that of etched titanium implants.

In addition to bridges, crowns and implants, other dental products are made from zirconium oxide ceramics. These include, for example, dental drills with three-edged geometry, which are very sharp and subject to virtually no wear.

Further Reading

1. Pospiech, P., Tinschert, J., & Raigrodski, A. (2004). Keramik–Vollkeramik. Ein Kompendium für die keramikgerechte Anwendung vollkeramischer Systeme in der Zahnmedizin. http://multimedia.3m.com/mws/media/598797O/lava-keramik-vollkeramik-kompendium.pdf?fn=Lava_Vollkeramik_Kompend_D.pdf. Accessed: 21. Nov. 2018.
2. Pospiech, P. (2014). Materialien für CAD/CAM-Technik: Die Qual der Wahl. https://www.zmk-aktuell.de/fachgebiete/digitale-praxis/story/materialien-fuer-die-cadcam-technik-die-qual-der-wahl__1047.html. Accessed: 28. Nov. 2018.
3. Sader, R., Lorenz, J., Holländer, J., & Ghanaati, S. (2015). Keramikimplantate—eine Übersicht. https://www.zwp-online.info/fachgebiete/implantologie/keramikimplantate/keramikimplantate-eine-uebersicht. Accessed: 22. Nov. 2018.
4. Tinschert, J., & Natt, G. (Hrsg.). (2007). *Oxidkeramiken und CAD/CAM-Technologien* (S. 5–22). Köln: Deutscher Ärzte Verlag.
5. DIN EN ISO 13356. (2016). Chirurgische Implantate—Keramische Werkstoffe aus yttriumstabilisiertem tetragonalem Zirkoniumoxid (Y-TZP).

Zirconia: A Synthetic Gemstone

21

Zirconia is a well-known and widely used imitation diamond for jewellery. It is a synthetically produced single crystal of zirconium oxide that has been stabilized in its high-temperature cubic modification (Sect. 14.2 in Chap. 14). Until the end of the 1970s, however, cultivated zirconium silicate, i.e. artificial zircon, was still used as a diamond imitate.

In nature, zirconium oxide occurs as the mineral baddeleyite (Chap. 11). Unlike natural zircon, baddeleyite does not form beautiful crystals worthy of cutting that can be used as gemstones. Baddeleyite also does not have a cubic (like zirconia), but a monoclinic crystal lattice. However, a natural cubic zirconia does exist and was first discovered around 1937 by two German mineralogists, M. F. von Stackelberg and K. Chudoba, in the form of microcrystals in metamorphic zircon.

In three countries, attempts have been made to produce zirconium oxide synthetically as a single crystal. This is why the gemstone has been given three names. In the experiments, the stabilisation of zirconium oxide in its cubic crystal structure, which has already been developed, was applied (Sect. 13.2 in Chap. 13). For this purpose, the yttrium oxide (Y_2O_3) or calcium oxide (CaO) is added during the production of zirconia. The correct designation for zirconia should actually be in the abbreviation: "CSZ stone", i.e. cubically stabilised zirconium oxide stone.

Zirconia was first synthesized in Russia in the early 1970s. At the Physical Institute of the Academy of Sciences (abbreviated to "FIAN"), it was possible to grow crystals of cubic zirconia that could also be ground. The zirconium oxide was stabilized with yttrium oxide. Under the name "Fianit"—derived from the name of the institute—these crystals were introduced to the market as gemstones. At about the same time, Americans also used the stabilization method with yttrium oxide. The crystals produced were named "Zirconia" by the company "Ceres Corporation". In 1976, the Swiss company Hrand Djevahirdji stabilized the zirconium oxide with calcium oxide and gave the artificial stones the name "Djevalith". As is known, only the name "Zirconia" has prevailed and is used today.

The production of cubic crystals from zirconium oxide requires special methods. Due to its very high melting temperature of approximately 2700 °C, no typical

© Springer-Verlag GmbH Germany, part of Springer Nature 2022
B. Arnold, *Zircon, Zirconium, Zirconia - Similar Names, Different Materials*,
https://doi.org/10.1007/978-3-662-64269-6_21

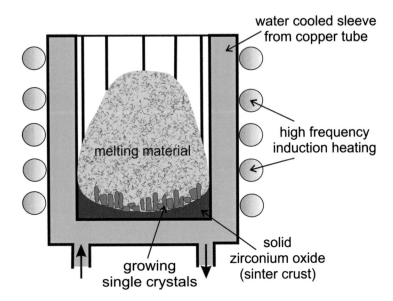

Fig. 21.1 Cold crucible process. (schematic)

crucibles can be used for growing from the melt, as no crucible material can survive this temperature. Therefore, a process known as the cold crucible method (skull-melting method) was developed and is most commonly used today. The process is shown schematically in Fig. 21.1.

The "crucible" in which the melt is produced consists of the material to be produced itself in the form of a solid crust which must first be formed. In this way, contamination by non-substance crucible substances can also be prevented. The starting material is filled as powder into a basket-like container made of water-cooled copper tubes (sleeve). This sleeve prevents possible melting of the crust.

The molten material is heated with the aid of a high-frequency induction heater. To couple the zirconium oxide, which is not metallic in itself, there is zirconium powder in the middle, which first heats up and then heats the oxide powder. As a good ion conductor, the zirconium oxide also couples at high temperature and a melt is formed. Due to the edge cooling, the oxide powder increasingly sintered inwards and independently forms an inherent, dense crucible, the "skull". Once the molten material has melted, the melting vessel is slowly pulled out of the induction coil, causing the temperature to drop and the actual crystallization to begin. Starting from nuclei on the crust, the elongated single crystals of zirconia grow into the melt and consume it. The crystals are broken out and further processed, while the crucible remains for the next powder filling. However, it must then be preheated using a different method (e.g. via laser) so that the energy can be transferred to the molten material. This method can also be used to produce other high-melting oxide crystals.

Single-crystalline materials are of great importance in modern technology. A well-known example are silicon single crystals made of pure silicon, which are cut into thin slices and used as so-called wafers in chip production. The targeted cultivation of single crystals with a specific morphology and a defined defect state (doping) is used, for example, for high-temperature superconductors and proteins.

Cubic zirconium oxide can form solid solutions with many other elements. This can result in zirconia single crystals of different colors. The addition of cerium, for example, results in an orange to red coloration. With nickel brown, with chromium green and with calcium blue shades are produced. For lavender-like colorations, neodymium oxide is added. However, usually only the colorless variety of zirconia is made into jewelry. Zirconia is made in all sizes and shapes and even with artificial inclusions. Because of these inclusions, even experts cannot distinguish good zirconia stones from diamonds after a visual inspection but only after suitable measurements (Chap. 22).

Certain manufacturers offer zirconia coated with DLC (Diamond like carbon). A DLC coating consists of diamond-like, amorphous carbon layers and increases the hardness of the stone. It also makes the coefficient of friction extremely low. An ADT (Amorphous diamond treatment) coating can also be applied using the CVD (Chemical vapour deposition) process. Only thick coatings are recognizable by interference phenomena on the surface.

Zirconia was classified as an artificial stone in the nomenclature until 1990. Since 1990, however, it has been declared a synthetic stone because it differs from natural zirconia, the mineral baddeleyite. Baddeleyite has a monoclinic crystal structure, whereas zirconia has a cubic structure. It also has a different composition due to stabilization with yttrium oxide or with calcium oxide. Thus zirconia is clearly a synthetic and not an artificial gemstone, i.e. a stone without a natural model. Artificial gemstones are identical in composition, properties and crystal structure to their natural representatives. Examples are ruby and sapphire, varieties of the natural aluminum oxide (corundum), which are often produced artificially.

Zirconia is not a genuine gemstone, but a cheap jewelry stone. However, due to its hardness and brilliance, and due to its advanced production process, zirconia is one of the most asked stones in the jewelry industry.

Figure 21.2 shows earrings made of zirconia. These earrings cost less than 30 EUR at the final seller, the price being determined more by the setting of rhodium-plated silver. Let's imagine what earrings of this size made of diamond would cost...

The mineral zircon (Chap. 3) is a genuine gemstone and is also used in the jewelry industry. However, zircon is considerably more expensive than zirconia and has some negative properties, so it has been almost completely displaced by zirconia stones. Zircon is not produced as a synthetic gemstone, but chemically produced zirconium silicate is used as a ceramic pigment, among other things.

The properties of zirconia are presented in the next chapter (Chap. 22) and are listed there in Table 22.1 in Chap. 22. Due to its cubic crystal structure and high hardness, zirconia is similar to diamond in many respects. In the brilliant cut, zirconia also comes very close to the diamond in terms of radiance. The similarities and also differences between zirconia and diamond will also be discussed in the next

Fig. 21.2 Zirconia earrings

chapter. Zirconia is still considered the best diamond imitation. However, since synthetic moissanite (silicon carbide) has been produced, it has a competitor.

Due to its optical properties, yttrium-stabilized zirconia (YCZ: yttrium cubic zirconia) is also used in technical applications. For example, lenses and laser elements are made from the material. Especially in the chemical industry, very resistant windows made of zirconia are used for the observation of highly corrosive liquids. Furthermore, YCZ is used as a substrate for semiconducting and superconducting films in electrical engineering.

Further Reading

1. Markl, G. (2015). *Minerale und Gesteine* (S. 48). Berlin: Springer Spektrum.
2. Schumann, W. (2017). *Edelsteine und Schmucksteine* (S. 267–270). München: BLV Buchverlag.
3. Wehmeister, U., & Häger, T. (2005). *Edelsteine erkennen—Eigenschaften und Behandlung* (S. 81). Stuttgart: Rühle-Diebener-Verlag.
4. Wikipedia. Cubic zirconia. https://en.wikipedia.org/wiki/Cubic_zirconia. Accessed: 11. Dez. 2018.
5. Rössler, L. Zirkonia. http://www.beyars.com/edelstein-knigge/lexikon_571.html. Accessed: 15. Dez. 2018.
6. Schorn, S. Mineralienatlas—Fossilienatlas. https://www.mineralienatlas.de/lexikon/index.php/K%C3%BCnstliche%20Kristalle. Accessed: 4. Dez. 2018.
7. Gold- und Platinschmiede Gerhards. Zirkonia. http://www.gold-platin-schmiede.de/wissenswertes/zirkonia/index.html. Accessed: 7. Dez. 2018.

Two Doubles of the Diamond

The natural mineral zircon and the synthetic zirconia (i.e. zirconium oxide) are used as diamond imitations in the jewellery industry—zircon (Chap. 3) since ancient times, zirconia (Chap. 21) since its first production in the twentieth century.

In brilliant cut, zircon (Fig. 1.1a in Chap. 1) is very similar to diamond, it looks almost the same. The cut zirconia (Fig. 21.2 in Chap. 21) is just as close in appearance to the diamond. Even an expert would have great difficulty in distinguishing these three stones merely by their appearance. Despite the outward similarity, however, we are dealing here with three very different gemstones. The diamond is a natural, very beautiful and very valuable gemstone. The zircon is also a natural, beautiful and valuable gemstone. The zirconia, although very beautiful, is a synthetic and rather worthless gemstone or, rather, jewelry stone.

What properties must a mineral, generally a material, possess to be considered a gemstone? The most important properties are high hardness (higher than 6 according to the Mohs scale), transparency and high refractive index and, if necessary, a beautiful color. In addition, the crystals of the material must be easy to cut. Table 22.1 lists the properties of the three gemstones considered and evaluated here.

Using the information in Table 22.1, we can now compare these three gemstones to find similarities as well as differences between them. Especially different properties are interesting, because with their help the stones can possibly be distinguished.

Chemically, the three gemstones are each individually composed (Table 22.1). Diamond consists of pure carbon. As we already know, zircon and zirconia have different compositions, despite the fact that they are related by name. Common to both stones is the contained chemical element zirconium. While zircon is a silicate ($ZrSiO_4$), zirconia is an oxide, more precisely zirconium dioxide (ZrO_2).

The mineral zircon crystallizes in the tetragonal crystal system. Zircon crystals form four-sided prisms with pyramids on the crystal ends. Zirconia, on the other hand, crystallizes in the cubic crystal system—like the model diamond—which is why it is sometimes called "cubic zirconia". Accordingly, the crystals of zirconia and diamond have the shape of cubes or double pyramids. Based on its crystalline

© Springer-Verlag GmbH Germany, part of Springer Nature 2022
B. Arnold, *Zircon, Zirconium, Zirconia - Similar Names, Different Materials*,
https://doi.org/10.1007/978-3-662-64269-6_22

Table 22.1 Properties of zircon, zirconia and diamond

Property	Zircon	Zirconia	Diamond
Origin	Natural	Synthetic	Natural artificial, if necessary
Chemical formula	$ZrSiO_4$	ZrO_2	C crystalline
Crystal system	Tetragonal	Cubic	Cubic
Density in g/cm^3	4.7	5.5	3.5
Hardness (Mohs scale)	7.5	8.5	10
Transparency	Transparent	Transparent	Transparent
Refraction	1.92 ... 1.99	2.2	2.417
Max. Birefringence	Strong 0.055	None	None
BG dispersion	0.038	0.066	0.044
Pleochroism	Weak to clear	No	No
Sheen	Diamond like	Glassy to diamond-like	Diamond sheen
Fracture	Conchoidal, very brittle	Conchoidal	Conchoidal to splintery
Fissility	Unclear	Non fissile	Completely
Thermal conductivity in W/m K	14[a]	3	2300

[a]Estimated after a comparative measurement with a diamond and gemstone tester

structure, zircon can be distinguished from the other two gemstones (Table 22.1). However, this structural examination requires special apparatus. It is not possible to distinguish between zirconia and diamond on the basis of structure. In this respect, a confusion of zirconia with diamonds is very likely.

The three gemstones have different densities (Table 22.1) and can thus be differentiated. Diamond and zircon are clearly lighter than zirconia. However, set stones are not suitable for determining the density.

We can see clear differences in the hardness values (Table 22.1). The hardest mineral in the world is and remains the diamond. It can only be cut with the help of other diamonds and with laser light. And this only works because the stone is completely cleavable and has different hardness values due to its anisotropy. In practice, however, the hardness can hardly be used to distinguish between gemstones, as hardness measurement is not a non-destructive test.

An important role in the evaluation and differentiation of gemstones is played by their optical properties. However, the transparency, i.e. the light transmission, of zircon, zirconia and diamond (Table 22.1) is the same and can therefore hardly be used to distinguish between them.

A very important characteristic value is the refractive index. The brilliance of the stone, i.e. the amount of light reflected from the interior of a gemstone, depends on this value. Diamond has the highest refractive index of the three gemstones discussed here (Table 22.1). At the same time, the equally high refractive index of zircon is noteworthy. The three gemstones can be distinguished by determining the refractive index. This determination is relatively simple and non-destructive.

Only zircons can be identified by birefringence (Table 22.1). Zircon has a tetragonal crystal lattice and is therefore birefringent. Diamond and zirconia are not birefringent due to their cubic crystal structure. Thus, they can be distinguished from zircon fairly easily with a magnifying glass.

Dispersion (light scattering) indicates the intensity of the colours produced when white light is split into minerals. Far more commonly, the term "fire" is used for color intensity. The three gemstones under consideration are characterized by unequal dispersion values (Table 22.1). Thus zirconia has a greater dispersion than zircon and even than diamond. Thus, by determining the dispersion, one can easily distinguish the three gemstones. However, the fire of gemstones can be increased by proper cutting, which optimally disperses the incident light. Incidentally, the so-called fire in diamond and zirconia is produced by the total reflection of light and not by birefringence. In zirconia, it is precisely the high birefringence that dampens its luminosity compared to diamonds.

Pleochroism is the property of birefringent crystals to split light into different colours in several directions (viewing angles). Consequently, this phenomenon can only be observed in zircon (Table 22.1) and then used for its determination.

The brilliance of a gemstone is not a measurable property and can only be described in words. Diamond as well as zircon and zirconia are intensely shiny stones. Similarly, the appearance of the fracture and cleavage are not measurable. Nevertheless, such differences between gemstones can be used to determine them.

The thermal conductivity is a suitable property for differentiation (Table 22.1). Diamond has the best thermal conductivity of all gemstones and also of all materials known today (with the exception of the very rare graphene). This means that it can be easily distinguished from zircon and zirconia with commercially available diamond gemstone testers that measure thermal conductivity, even when prepared and even with the smallest stones.

For the differentiation of gemstones, other characteristics can be used in addition to those mentioned above. These include, in particular, inclusions, which are often referred to as "nature's certificate of authenticity". However, in order to come as close as possible to nature, zirconia crystals are even produced with inclusions.

So the above comparison of properties shows us that the two materials zircon (zirconium silicate) and zirconia (zirconium oxide) are very similar to the diamond. This justifies their designation as "doubles of the diamond".

Further Reading

1. Schumann, W. (2017). *Edelsteine und Schmucksteine* (S. 274–275). München: BLV Buchverlag.
2. Purle, T. acquimedia. Zirkon und Zirkonia unterscheiden. http://www.steine-und-minerale.de/artikel.php?topic=1&ID=159. Accessed: 11. März 2019.

Index

© Springer-Verlag GmbH Germany, part of Springer Nature 2022
B. Arnold, *Zircon, Zirconium, Zirconia - Similar Names, Different Materials*,
https://doi.org/10.1007/978-3-662-64269-6

Printed in the United States
by Baker & Taylor Publisher Services